CRITICAL THINKING & LOGICAL REASONING PRIMER

P

GIFT OF LOGIC™ SERIES

Boost Your Thinking Skills

An Essential Resource for Everyone

Verbal Reasoning
Analytical Reasoning
Pictorial Reasoning

THIRD EDITION

| FOR GRADES 6-12 | STUDENTS, TEACHERS, AND PARENTS |

Prerequisite for Workbooks 6-10

Ranga Raghuram   **GIFT OF LOGIC™**

Gift Of Logic, Inc
http://www.giftoflogic.com
sales@giftoflogic.com

Critical Thinking and Logical Reasoning Primer
ISBN-13: 978-1494832360
ISBN-10: 1494832364

Third Edition
1-2014

Copyright © 2009 Gift Of Logic, Inc. All rights reserved. No part of this publication may be reproduced, stored in a retrieval system, transmitted in any form or by any means, electronic, mechanical, photocopying, recording or otherwise, without the written permission of the publisher.

License: This book is licensed for use by one person only. Use of this book in a group setting (classroom, workshop, etc) without the written permission of the publisher is prohibited. Unauthorized duplication is strictly prohibited by law. Contact the publisher at sales@giftoflogic.com for classroom/school/group licensing.

GIFT OF LOGIC™
CRITICAL THINKING & LOGICAL REASONING CURRICULUM
12 WORKBOOKS TO BOOST YOUR THINKING SKILLS

For Kindergarten, Grade 1, and Grade 2

Workbook# 0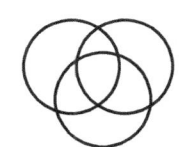

Verbal Reasoning	Finding the truth, Inferencing, Analogies, Synonyms and Antonyms, Agree/Disagree
Analytic Reasoning	Memory drill, Decision making, Positioning, Sudoku
Pictorial Reasoning	Connect the dots, Mazes, Picture Sequence, Spot the difference, etc

Workbook# 1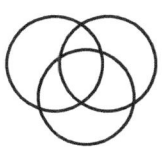

Verbal Reasoning	Finding the truth, Inferencing, Analogies, Synonyms and Antonyms, Agree/Disagree
Analytic Reasoning	Sorting, Positioning, Picking, Assorted problems, Numeric and Alphabetic Sudoku
Pictorial Reasoning	Picture Sequence, Spot the difference, Odd picture

Workbook# 2

Verbal Reasoning	Finding the truth, Classification, Direct and Inverse relationship, Inferencing, Analogies, Agree/Disagree
Analytic Reasoning	Sequencing, Scheduling, Strategy, Picking, etc
Pictorial Reasoning	Picture Analogy, Odd picture, Pattern matching, etc

For Grade 3, Grade 4, and Grade 5

Workbook# 3

Verbal Reasoning	Not, And, Or, If .. then, Conditional inferencing, Unconditional inferencing, Symbolic Logic
Analytic Reasoning	Lists, Sequencing, Grouping, Venn Diagrams, Graph logic, Number logic, Letter logic, Sudoku
Pictorial Reasoning	Picture sequence, Picture analogy, Odd picture, Picture difference, Pattern matching

Workbook# 4

Verbal Reasoning	Contradiction, Converse, Inverse, Contrapositive, Conditional inferencing, Symbolic Logic
Analytic Reasoning	Scheduling, Looping, FIFO, LIFO, Correlation, Venn Diagram, Graph logic, Number logic, Sudoku, etc
Pictorial Reasoning	Picture sequence, Picture analogy, Odd picture, Picture difference, Pattern matching

Workbook# 5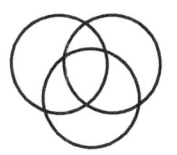

Verbal Reasoning	Biconditional, Categorical inferencing, Cause and Effect, Symbolic Logic, Agree/Disagree, Word and Sentence analogy
Analytic Reasoning	Correlation, Grouping, Venn Diagrams, Graph logic, Number logic, Letter logic, Sudoku, etc
Pictorial Reasoning	Picture sequence, Picture analogy, Odd picture, Picture difference, Pattern matching

********* Essential resource for everyone *********

*http://www.giftoflogic.com *sales@giftoflogic.com

GIFT OF LOGIC™
CRITICAL THINKING & LOGICAL REASONING CURRICULUM
12 WORKBOOKS TO BOOST YOUR THINKING SKILLS

For Grades 6-12, College/University Students, Adults

Primer

Prereq

Verbal Reasoning	Logical Operators, Conditional, Categorical and Causal reasoning, Validity, Fallacies, Symbolic Logic
Analytic Reasoning	Positioning, Grouping, Sudoku
Pictorial Reasoning	Pattern perception, Figure formation, Paper folding and cutting, Figure matrix, Rule detection

Workbook# 6

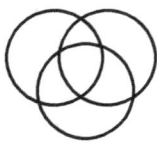

Verbal Reasoning	Arguments-Main point, Must be true, Cannot be true
Analytic Reasoning	Positioning, Grouping, Sudoku
Pictorial Reasoning	Pattern perception, Figure formation, Paper folding and cutting, Figure matrix, Rule detection

Workbook# 7

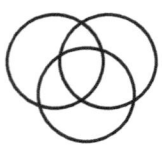

Verbal Reasoning	Arguments-Strengthening, Weakening
Analytic Reasoning	Positioning, Grouping, Sudoku
Pictorial Reasoning	Pattern perception, Figure formation, Paper folding and cutting, Figure matrix, Rule detection

Workbook# 8

Verbal Reasoning	Arguments - Controversy, Paradox
Analytic Reasoning	Positioning, Grouping, Sudoku
Pictorial Reasoning	Pattern perception, Figure formation, Paper folding and cutting, Figure matrix, Rule detection

Workbook# 9

Verbal Reasoning	Arguments- Assumptions, Reasoning strategy
Analytic Reasoning	Positioning, Grouping, Sudoku
Pictorial Reasoning	Pattern perception, Figure formation, Paper folding and cutting, Figure matrix, Rule detection

Workbook# 10

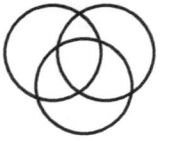

Verbal Reasoning	Arguments-Flawed reasoning, Analogous reasoning
Analytic Reasoning	Positioning, Grouping, Sudoku
Pictorial Reasoning	Pattern perception, Figure formation, Paper folding and cutting, Figure matrix, Rule detection

********* Essential resource for everyone *********
Get the GIFT OF LOGIC™ today !
*http://www.giftoflogic.com *sales@giftoflogic.com

Dear Reader:

Your decision to purchase this book is commendable. You now have in your hands, a comprehensive book in Critical thinking and Logical reasoning. This primer covers several essential topics in Verbal, Analytical and Pictorial reasoning. This book is an essential prerequisite for Workbooks 6-10 in the Gift Of Logic(tm) -Critical thinking & Logical reasoning series.

Topics in Verbal Reasoning cover both deductive (valid/invalid) and inductive arguments. Topics such as logical operators, argument structure and deduction/inferencing are covered in detail. Conditional reasoning, categorical reasoning and causal reasoning are heavily emphasized as they are frequently encountered in day to day life. You will also learn to identify various types of flawed arguments.

Topics covered in Analytical Reasoning include positioning and grouping problems and how to represent scenarios using symbols and solve problems quickly.

Topics in Pictorial Reasoning include pattern perception, figure formation, paper folding and cutting, figure matrix and rule detection.

This book is one in a series of twelve workbooks. This book provides the essential foundation for Workbooks 6-10. After reading this book, continue to do the exercises in Workbooks 6-10. Please refer to the brochure before this page for a brief description of each workbook. Visit the website http://wwww.giftoflogic.com for more information.

                Happy thinking and reasoning !

TABLE OF CONTENTS

Verbal Reasoning

Finding the truth...9
Logical operators..10
Negation..12
Conjunction...13
Disjunction..14
Arguments structure...16
Deductive and Inductive arguments..20
Conditional arguments...25
Categorical arguments...45
Other deductive arguments..55
Causal arguments...56
Statistical arguments..63
Analogous arguments.. 64
Flawed arguments..66
Further Reading...82

Analytic Reasoning

Positioning...85
Grouping..94

TABLE OF CONTENTS

Pictorial Reasoning

Problem description...99
Connect the dots..101
Maze..102
Picture Sequence 103
Odd Picture...104
Picture Difference...105
Picture Analogy..106
Pattern Matching..107
Pattern Perception - Missing pattern....................................108
Pattern Perception - Continuing pattern...............................109
Figure Formation..110
Paper folding and cutting...111
Figure Matrix - Similarity..112
Figure Matrix - Analogy..113
Rule Detection..114

Certificate of Completion

VERBAL REASONING

FINDING THE TRUTH

Finding the truth of a statement is important in Critical thinking and Logical reasoning. Some statements are true or false without any doubt. But, some statements cannot be proven to be either true or false. Such statements are called uncertain statements.

Example: Monkeys cannot climb trees.
Since we know for a fact that monkeys can climb trees, this statement is false without any doubt.

Example: Five years from now, human beings will land in Mars.
We cannot be sure that human beings will land in Mars five years from now. They may or may not. So, this statement is uncertain.

Find the truth of the following statements and circle the correct answer.

- A duck is not a mammal.

    * True   * False   * Uncertain

- Paris is the capital of England.

    * True   * False   * Uncertain

- It will rain tomorrow.

    * True   * False   * Uncertain

Verbal Reasoning

LOGICAL OPERATORS

The following table shows the logical operators and their symbols.

Operator	Symbol	Description	Expressed using
not	~	Negation	not it is not the case that
and	&	Conjunction	and both
or	∥ (inclusive) ╫ (exclusive)	Disjunction	or, either .. or either .. or but not both
if	→	Conditional	if, if … then, only if, when, unless, except, without
if and only if	↔	Biconditional	if and only if

These operators must be understood clearly, since they play a very important role in Critical thinking and Logical reasoning.

Next, we will learn about these logical operators in detail.

Verbal Reasoning
© Gift Of Logic, Inc * Copying prohibited

LOGICAL OPERATORS

Operator	Examples
not	Washington DC is <u>not</u> the capital of China. Paris is <u>not</u> the capital of France. It is <u>not</u> the case that human beings can live forever. It is false to say that the Sun is <u>not</u> at the center of the Solar system.
and	He is the best soccer player <u>and</u> the best baseball player. She is a good singer <u>and</u> a good athlete.
or	You may do the homework in the morning <u>or</u> in the evening. You can take either the bus <u>or</u> the train to college.
if	<u>If</u> you paint the house, it will look clean. <u>If</u> Rosie bakes a cake, we will eat it. <u>When</u> the clock strikes ten, the store will open. <u>Except</u> this tire, all other tires are flat. <u>Unless</u> you try hard, you are not going to win.
if and only if	The Mayor will come <u>if and only if</u> the Governor also comes. Cindy will sing if and only if Selena also sings.

TRUTH TABLE FOR NOT (Negation)

P	~P
true	false
false	true

Truth tables show the different values a logical operator can take.
The above table is the truth table for logical operator "not".
P is a statement that can be true or false.
Note the symbolic notation. Negation of P is represented as ~P.

If P is true, ~P is false. If P is false, ~P is true.

Negation of a true statement is a false statement.
Negation of a false statement is a true statement.

  P: Earthquakes can destroy buildings. (P is true)
~P: Earthquakes cannot destroy buildings. (negation ~P is false)

  P: Digestion is the function of the heart. (P is false)
~P: Digestion is not the function of the heart (negation ~ P is true).

Verbal Reasoning
© Gift Of Logic, Inc * Copying prohibited

TRUTH TABLE FOR AND (Conjunction)

P	Q	P & Q
True	True	True
True	False	False
False	True	False
False	False	False

The above table is the truth table for the logical operator "and".

Conjunction of two statements P and Q is represented as P & Q. P & Q will be true if both P and Q are separately true. If one of the statements is false or if both are false, the conjunction will be false. P and Q are called "conjuncts".

True and True example:
    We see with our eyes <u>and</u> we hear with our ears.
        true and true = true

True and False example:
    We see with our eyes <u>and</u> we hear with our nose.
        true and false = false

False and True example:
    We see with our nose <u>and</u> we hear with our ears.
        false and true = false

False and False example:
    We see with our nose <u>and</u> we hear with our eyes.
        false and false = false

TRUTH TABLE FOR OR (Disjunction)

P	Q	P ‖ Q
True	True	True
True	False	True
False	True	True
False	False	False

The above table is the truth table for the logical operator "or". Disjunction of two statements, represented as P ‖ Q, is true if just one of them is true. P and Q are called "disjuncts".

True or True example:
    We see with our eyes or we hear with our ears.
        true or true = true

True or False example:
    We see with our eyes or we hear with our mouth.
        true or false = true (since one of them is true)

False or True example:
    We hear with our eyes or we smell with our nose.
        false or true = true (since one of them is true)

False or False example:
    We hear with our eyes or we smell with our mouth.
        false or false = false

| INCLUSIVE OR ‖ | EXCLUSIVE OR ╫ |

The logical operator "or" can be used in an inclusive context or an exclusive context. Inclusive Or is represented by symbol ‖. Exclusive Or is represented by symbol ╫.

Inclusive Or: P ‖ Q. This means either P or Q or both. Think of the two lines in the symbol as representing each option. Since it is not crossed, "both" is also an option.

Exclusive Or: P ╫ Q. This means either P or Q, but not both. Think of the two lines in the symbol as representing each option. Since the symbol is crossed, "both" is not an option.

Example: We can drink milk or water. Symbolic: milk ‖ water
In this statement, it is possible that we can drink both. So the "or" is used in the inclusive context.

Example: He is either upstairs or downstairs now. upstairs ╫ downstairs
In this statement, the "or" is used in an exclusive context because one can be at only one place at any given moment.

To communicate clearly without any doubt, it is better to use the phrase "but not both".

Example: You can drink juice or soda. (juice ‖ soda)
You can drink juice or soda, but not both. (juice ╫ soda)

Discussion about the "if" and "if and only if" operators are presented in the context of argumentation.

ARGUMENT STRUCTURE

Argument: An argument typically consists of one or more premises and one conclusion. Occasionally, it may consist of one or more premises, one or more sub-conclusions, and a final conclusion. It is also possible that an argument is made of just one sentence that has one or more premises and a conclusion. The premises provide the facts that lead to the conclusion.

Why do we need arguments? Arguments are needed to prove a point, resolve a controversy, reach a conclusion, and so on. But, since the premises in an argument can be presented in several ways and the conclusion in the argument can be derived in several ways, it is very important to identify a valid argument from an invalid argument. By studying argument structure and understanding deductive reasoning, we can tell a valid argument from an invalid one. The method by which we use the premises to reach the conclusion is called reasoning. The result of reasoning is expressed in the conclusion.

Example of a simple argument:
    All birds can fly. A parrot is a bird. Therefore, a parrot can fly.

The premises of the argument are:
    All birds can fly.
    A parrot is a bird.

ARGUMENT STRUCTURE

The conclusion of the argument is:
   Therefore, a parrot can fly.

The conclusion indicator is the word "Therefore".

This argument has two premises, but other arguments can have several premises. The conclusion can appear in the beginning or the end of the argument. Intermediate conclusions can appear in the middle of the argument.

Certain words clearly indicate the presence of premises and counter premises in the argument. Certain words clearly indicate the conclusion in the argument.

Premise: Premises are facts that support the conclusion. Premises must be such that they should lead us to the correct conclusion. The following is a list of words/phrases commonly used to indicate premises in arguments:

   because, since, granted that, moreover, besides

Statements that do not have premise indicators, but contain facts that are used to support the conclusion are also premises for the argument.

ARGUMENT STRUCTURE

Examples of premise statements:
1) The weather will be warm today.
2) Washington DC is the capital of USA.
3) The geography quiz was easy. Moreover, we had to answer only ten questions.
4) He is very tall.

Counter Premise: A counter premise is used to contradict an earlier premise. Words that are commonly used as counter premise are:

but, however, nevertheless, regardless, although, even though, whereas

Examples of counter premise statements:

1) You must water the plants now. However, do not water more than necessary.
2) These homes are expensive whereas those homes are cheap.
3) Although today is a holiday, some shops are open.
4) I do not like to travel. Nevertheless, I will take a flight to come and see you next week.
5) This patient is in serious condition. But, he is expected to recover soon.

Verbal Reasoning

ARGUMENT STRUCTURE

Conclusion: The conclusion of an argument can be in the beginning or in the end of the argument. Conclusion is normally identified by conclusion indicator words and phrases such as the following:

therefore, thus, hence, in conclusion,
for this reason, consequently, infer

Example of conclusion statements:
1) So, I have decided to include you in my team.
2) Therefore, we must stop wasting water.
3) For this reason, I declare an emergency for our state.
4) In conclusion, we must pledge to support our troops.
5) Consequently, we must run to a shelter immediately.

English language is so flexible that you can sometimes find premises and conclusion in the same sentence. The following are one sentence arguments. The conclusion is underlined.

One sentence arguments:

1) Since there are a lot of flowers in the garden, <u>bees are attracted to it</u>.
2) War is devastating and so <u>it must be avoided</u>.
3) <u>I woke up late today</u> because I watched a movie last night.
4) I like sugar and that is the reason <u>I eat a lot of ice cream</u>.

Verbal Reasoning
© Gift Of Logic, Inc * Copying prohibited

DEDUCTIVE & INDUCTIVE ARGUMENTS

Arguments are classified as **Deductive or Inductive (but not both)**.

In a deductive argument, you can tell if the conclusion is true or false. In an inductive argument, you cannot tell if the conclusion is true or false. It is uncertain. It maybe true or maybe false.

A **valid argument** is an argument in which, if the premises are true, then the conclusion is also true. That is, the conclusion can be deduced/inferred from the premises.

An **invalid argument** is an argument in which, if the premises are true, the conclusion is false. That is, the conclusion cannot be deduced/inferred from the premises.

Examples of valid arguments:

London is in England.
England is in Europe.
Therefore, London is in Europe.

All birds have feathers.
Eagles are birds.
Therefore, eagles have feathers.

Note that when we say that an argument is valid or invalid, we are saying that the reasoning in the argument is valid or invalid.

Verbal Reasoning

DEDUCTIVE & INDUCTIVE ARGUMENTS

Example of an invalid argument:
>   The eye is in the head.
>   The nose is in the head.
>   Therefore, the eye is in the nose.

The test for validity is "If the premises are true, is the conclusion also true?". The conclusion is that the eye is in the nose, which is clearly false. This is an example of an invalid argument. This argument is invalid because it assumes that if two things are found in the head, then one must be in the other. The two things may be next to each other instead of being inside each other. The reasoning in the argument is invalid. The conclusion cannot be deduced/inferred from the premises.

False premises can lead to valid arguments.

>   Paris is the capital of India.
>   India is in Asia.
>   Therefore, Paris is in Asia.

Is this argument valid or an invalid? It is valid because, if the premises are true, then the conclusion is also true. The fact that one of the premises is false is not important for determining validity. You have to assume that the premises are true (even if they really are not) and test if the conclusion is true or false. If Paris is the capital of India and if India is in Asia, then we can infer that Paris is in Asia. This argument is valid. But, we know that the first premise (Paris is the capital of India) is false. The second premise (India is in Asia) is true.

Verbal Reasoning

DEDUCTIVE & INDUCTIVE ARGUMENTS

The conclusion (Paris is in Asia) is false. But, the argument is valid! If the premises are true, then the conclusion is true. The reasoning behind the argument is valid when in fact one of the premises and the conclusion itself is false.

Just because an argument is valid does not mean that its conclusion must be true, like the example shown above. Everyone knows that Paris is not in Asia. So, this argument, while valid, is not a **sound** argument.

A **sound argument** is one which is a) valid and b) whose premises are true. So, in the above argument, the argument is valid, but one of the premises is not true. Therefore, the argument is not sound.

So, it does not make sense to make arguments that are not sound. But, people make unsound arguments anyway to convince or prove their point of view. Therefore, it is important to learn to identify whether the argument is sound or not. When you get cheated by someone, it is very likely that you were convinced by a valid argument that is not sound!

In practice, it is hard to communicate to everyone the difference between a valid argument and a sound argument. Most people will understand and speak of only valid and invalid arguments. Further discussion will focus on whether an argument is valid or invalid only. Remember that the test of validity is done using the following logic:

> If the premises are true, is the conclusion also true?
>> If the answer is yes, then it is a valid argument.
>> If the answer is no, then it is an invalid argument.

Verbal Reasoning

DEDUCTIVE & INDUCTIVE ARGUMENTS

Inductive arguments are those whose conclusion cannot be deduced (inferred) with certainty. That is, we cannot say if the conclusion is true or false. It may be true or may be false. The reasoning in inductive arguments can be strengthened or weakened by means of additional facts (premises).

Example of an Inductive Argument:
   Eighty percent of men like to drive fast.
   Neil is a man.
   Therefore, Neil likes to drive fast.

Even if the premises are true, we cannot deduce with certainty whether the conclusion will be true or false. It maybe true or maybe false. So, this is an inductive argument. If ninety percent of men liked to drive fast, then Neil is more probable to drive fast. If only twenty percent of men liked to drive fast, then Neil is less probable to drive fast. Thus, this inductive argument can be made stronger or weaker depending on the strength of the premises.

Whether an argument is deductive (valid/invalid) or inductive, there is reasoning involved in each of them. The process of reaching the conclusion from the premises is called reasoning. The result of reasoning is expressed in the conclusion of the argument. There are several types of reasoning. We will now learn about them in the context of conditional arguments, categorical arguments, causal arguments, etc. We will also discuss several types of flaws that are found in reasoning.

DEDUCTIVE & INDUCTIVE ARGUMENTS

Summary:

Arguments consist of premises and a conclusion. Premises provide the facts that lead to the conclusion. The process of reaching the conclusion is called reasoning. The result of reasoning is expressed in the conclusion.

Arguments can be Deductive or Inductive. Deductive arguments can be valid or invalid. To find out if an argument is valid, apply the following rule:

> If the premises are true, is the conclusion also true?
> If the answer is yes, then it is a valid argument.
> If the answer is no, then it is an invalid argument.

Just because an argument is valid does not mean that it is a sound argument. A sound argument is a valid argument whose premises are true.

There are several ways to state the premises and conclusion within an argument. It is important to spot the premises and conclusion in an argument, and evaluate whether the reasoning in the argument is valid or invalid.

Next, we will discuss different types of arguments and learn the type of reasoning they use to reach their conclusion.

CONDITIONAL ARGUMENTS

*Conditional arguments have conditional premises and/or conditional conclusions in them.

*Conditional premises are stated using logical operators:
      if, if .. then, if and only if

*Conditional premises can also be stated using words such as:
      when, except, unless, without

*You should read carefully and learn to spot conditional statements.

Examples of conditional premises:
1) If Jack goes to the movie, then Jill will go to the movie.
2) Jack will go to the movie if Jill goes to the movie.
3) Jack will go to the movie, if and only if Jill goes to the movie.
4) Unless Jack comes, Jill will not come.
5) Without four inflated tires, the car will not run.

Note the position of the words "if" and "only if" in the above statements. The position is important, since it can change the meaning of the condition. Note that the above statements are not conclusions since conclusion keywords such as therefore, hence, etc are not used.

Examples of conditional conclusions:
    Therefore, if Jack goes, then Jill will not go.
    Hence, unless someone fixes this leak, a lot of water will be wasted.

Verbal Reasoning
© Gift Of Logic, Inc * Copying prohibited

CONDITIONAL ARGUMENTS

If conditions are present in the premises, we can make inferences (deductions) based on these conditions. From now on, premises in an argument are represented by letters P, Q, R etc. Recall that the logical operator "if" is represented symbolically using the symbol →.

Conditional premise: If P then Q
Symbolic: P → Q
P (antecedent) → Q (consequent)

The operand P on the left side is called the antecedent or sufficient condition or hypothesis. The operand Q on the right side is called the consequent or necessary condition or conclusion. You must learn to represent conditional statements in symbolic form.

Premise: If Jack goes to the movie, then Jill will go to the movie.
Symbolic: Jack goes to movie → Jill goes to movie

Since we know that the context of this premise is "going to the movie", we could also write this statement in a shorter symbolic form as:
Jack → Jill. It does not matter how descriptive we are to describe the antecedent and the consequent. What matters is that antecedent and the consequent are accurately identified and understood.

Please refer to workbook# 3, 4, and 5 for a rudimentary discussion of conditional, categorical, and causal statements and their representation in symbolic form.

CONDITIONAL ARGUMENTS — Conditional

Argument Structure	Symbolic	
If P then Q P Therefore Q	P → Q P ∴ Q	valid argument

∴ is the symbol used to represent "Therefore".

In the above argument structure, "If P then Q" is the conditional premise. The second premise states that P is true. The conclusion, "Therefore Q", means Q is true. If the antecedent occurs, then the consequent will occur. If the premises are true, then the conclusion is true. So, this argument is a valid argument.

Example:

If Adam runs fast, he will win the race. | Adam runs fast → win race
Adam ran fast. | Adam ran fast
Therefore, he won the race. | ∴ Adam won

In this example, if Adam runs fast, he will win the race without any doubt. From the condition alone, we do not know if he ran fast or not. This is clarified in the second premise which says that he ran fast. Therefore, we can infer that he won the race. This argument is valid. There are other variations of conditional arguments that we will discuss next.

CONDITIONAL ARGUMENTS — Contradiction

Argument Structure	Symbolic	
If P then Q P Therefore Not Q	P → Q P ∴ ~Q	contradiction is false invalid argument
If P then Q Not P Therefore Q	P → Q ~P ∴ Q	contradiction is false invalid argument

The contradiction of conditional P → Q is P → ~Q or ~P → Q. Since the contradiction violates the condition, it is clearly false. So, an argument whose conclusion is a contradiction of the conditional is an invalid argument.

Example:

If Paul runs fast, he will win the race. Paul ran fast. Therefore, he did not won the race.	Paul runs fast → win race Paul ran fast ∴ ~win race

If the condition is true, then since Paul ran fast (second premise), he should have won the race. To conclude that he did not win the race is a contradiction of the condition. So, the argument is invalid. In other words, if P then Q does not imply if P then Not Q (or) if Not P then Q.

CONDITIONAL ARGUMENTS — Converse (Reversal)

Argument Structure	Symbolic	
If P then Q Q Therefore P	$P \rightarrow Q$ Q $\therefore P$	converse(reversal) is false invalid argument

The converse (negation) of a conditional $P \rightarrow Q$ is $Q \rightarrow P$. If the condition is true, then the converse is false.

The condition is that if P occurs, Q will occur. If Q occurs, it is not necessary that P must occur. Q can occur whether or not P occurs, but if P occurs, then Q must occur. So, an argument whose conclusion is the converse of the conditional is an invalid argument.

Example:

If Adam runs fast, he will win the race. Adam won the race. Therefore, he ran fast.	Adam runs fast $\rightarrow$ win race Adam won race $\therefore$ Adam ran fast

This argument is invalid because it concludes that the converse of the condition is true. It is possible that Adam won the race even though he did not run fast because others ran slower than him.

In other words, "if P then Q" does not imply "if Q then P".

CONDITIONAL ARGUMENTS   Inverse (Negation)

Argument Structure	Symbolic	
If P then Q	P → Q	inverse(negation) is false
Not P	~P	
Therefore not Q	∴ ~Q	invalid argument

The inverse (negation) of a conditional P → Q is ~P → ~Q. If the condition is true, then the inverse is false.

The condition only says that if P happens, Q will happen. If P does not happen, then Q can still happen. So, an argument whose conclusion is the inverse (negation) of the conditional is an invalid argument.

Example:

If Paul runs fast, he will win the race.	Paul runs fast → win race
Paul did not run fast.	~run fast
Therefore, he did not win the race.	∴ ~win race

This argument is invalid because, even if Paul does not run fast, he could still win the race if others run slower than him. So, we cannot conclude for sure that if he does not run fast, he will not win the race. In other words, "if P then Q" does not imply "if Not P then Not Q".

CONDITIONAL ARGUMENTS — Contrapositive

Argument Structure	Symbolic	
If P then Q	P → Q	contrapositive is true
Not Q	~Q	
Therefore Not P	∴ ~P	valid argument

The contrapositive of a conditional P → Q is ~Q → ~P.
If the conditional is true, then the contrapositive is true.

If P happens, then Q will happen.
So, if Q did not happen, then P did not happen.

Example:

If Paul runs fast, he will win the race.	Paul runs fast → win race
Paul did not win the race.	~win race
Therefore, Paul did not run fast.	∴ ~Paul run fast

The condition is that if he ran fast, he will win the race. Since Paul did not win the race, we can infer that he did not run fast. This is a valid argument.

In other words, "if P then Q" implies that "if Not Q then Not P".

Verbal Reasoning

CONDITIONAL ARGUMENTS - SUMMARY

Argument Structure	Symbolic	Remarks
If P then Q P Therefore Q	$P \rightarrow Q$ P $\therefore Q$	conditional valid argument
If P then Q P Therefore Not Q	$P \rightarrow Q$ P $\therefore \sim Q$	contradiction is false. invalid argument
If P then Q Q Therefore P	$P \rightarrow Q$ Q $\therefore P$	converse (reversal) is false invalid argument
If P then Q Not P Therefore not Q	$P \rightarrow Q$ $\sim P$ $\therefore \sim Q$	inverse (negation) is false invalid argument
If P then Q Not Q Therefore Not P	$P \rightarrow Q$ $\sim Q$ $\therefore \sim P$	contrapositive is true valid argument

Next, we will see how the conditional operator "if" is used in other situations such as "only if", "if and only if" etc.

Verbal Reasoning

CONDITIONAL ARGUMENTS   P only if Q

Argument Structure	Symbolic	
P only if Q	P → Q	P only if Q is the same as If P then Q.
P	P	
Therefore Q	∴ Q	contrapositive ~Q → ~P is also valid

The statement "P only if Q" effectively means "If P then Q". Note how the positions of P and Q have changed in the "if .. then" statement.

Example: Consider the following "only if" statement:
    Only if: The mayor will come, only if the governor comes.

The logical possibilities are:
1) Mayor only comes – not possible – mayor will come only if the governor comes.
2) Governor only comes – possible - the governor can come without the mayor.
3) Mayor and Governor both come – possible - this satisfies the "only if" statement.

From the above, we can infer that effectively, the "only-if" statement can be translated to the following "if-then" statement:
 "If the mayor comes, then the governor will come".

Note the change in position of mayor and governor in the "if-then" statement. In symbolic form, we can write this as:

Verbal Reasoning

CONDITIONAL ARGUMENTS — Only if P then Q

Argument Structure	Symbolic	Only if P, then Q is the same as
Only if P then Q	Q → P	If Q then P
Q	Q	
Therefore P	∴ P	contrapositive ~P → ~Q is valid

The statement "Only if P then Q" effectively means "If Q then P". Note how the positions of P and Q have changed in the "if .. then" statement.

Whenever you see "only if", you can convert it to a "if then" by switching the operands P and Q.

Example:
    Only if the governor comes, will the mayor come.

This "only if" can be translated to the following conditional:
        If the mayor comes, then the governor will come.
           mayor comes → governor comes

Note that since the contrapositive of a conditional is a valid inference, we can conclude the following:

        If the governor does not come, the mayor will not come.
        ~governor come → ~mayor come

Verbal Reasoning

CONDITIONAL ARGUMENTS Biconditional If and only if P then Q

Argument Structure	Symbolic	
If and only if P then Q P Therefore Q	P ↔ Q (P → Q and Q → P) P ∴ Q	valid argument contrapositive ~Q → ~P is valid
If and only if P then Q Q Therefore P	P ↔ Q (P → Q and Q → P) Q ∴ P	valid argument contrapositive ~P → ~Q is valid

The statement "If and only if P then Q" is called a biconditional statement. It has two conditions in it. 1) If P then Q 2) only if P then Q (this is the same as if Q then P)

Combining these two, we can say that "If and only if P then Q" is the same as "If P then Q" and "If Q then P". This is symbolically represented as P ↔ Q. ↔ is the symbol for the biconditional operator.

Example: Sam will win the race if and only if he runs fast.
Sam will win ↔ runs fast

From this biconditional statement, we can infer that:
1) If Sam wins the race, he ran fast.
2) If Sam runs fast, he will win the race.

CONDITIONAL ARGUMENTS — conditional chaining

Argument Structure	Symbolic	
If P then Q	P → Q	valid argument
If Q then R	Q → R	
P	P	contrapositive ~R → ~P is valid
∴ R	∴ R	

Inferences can be drawn by logical chaining of several conditionals as long as we follow all the inferencing rules regarding contradiction, converse, inverse, and contrapositive statements.

Example: If you pay your tax, you obeyed the law.
If you obeyed the law, you are a good citizen.
Therefore, if you paid your tax, you are a good citizen.

The above argument is valid. The conclusions can be drawn from logical chaining.

> pay tax → obeyed law
> obeyed law → good citizen
> ∴ pay tax → good citizen
> ∴ ~good citizen → ~pay tax (valid contrapositive)

Verbal Reasoning

CONDITIONAL ARGUMENTS — conditional chaining

When the antecedent (P) or consequent (Q) involves negation, we should be careful in making the logical chain and drawing inferences.

P → Q ; Q → ~R
∴ P → ~R is a valid inference
∴ R → ~P is a valid inference (contrapositive)

Note that Q and Q chain together.

P → ~Q ; ~Q → ~R
∴ P → ~R is a valid inference
∴ R → ~P is a valid inference (contrapositive)

Note that ~Q and ~Q chain together.

P → ~Q ; Q → ~R
∴ P → ~R is a invalid inference

Note very carefully that in this case, a conditional chain does not exist. ~Q and Q do not chain together.

Verbal Reasoning

CONDITIONAL ARGUMENTS — Unless, Without

Conditions do not have to expressed only with "if .. then" statements. Words such as when, unless, until, without and except can also be used to express conditions. But, since making logical deduction is easier with "if .. then" statements, we can convert statements that do not have the "if.. then" form to statements with "if..then" form and then make our inferences.

Unless in the beginning

Unless P happens, Q will not happen.
       converts to
If P does not happen, Q will not happen.
    $\sim P \to \sim Q$
    $\therefore Q \to P$ (contrapositive)

Example: Unless we fix the tire, the car will not move.
       can be converted to
If we do not fix the tire, the car will not move.
If the car moves, then we have fixed the tire. (contrapositive inference)

Without in the middle

P will not happen without Q.
       converts to
If Q does not happen, P will not happen.
    $\sim Q \to \sim P$
    $\therefore P \to Q$ (contrapositive)

Example:
The circus will not begin without the elephant.
       can be converted to
If the elephant is not present, the circus will not begin.
Therefore, if the circus began, the elephant was present. (contrapositive)

Verbal Reasoning

CONDITIONAL ARGUMENTS — Except

Argument structure using Except	Symbolic	
All must satisfy P except Q converts to If Not Q then P Therefore, if Not P then Q	P except Q converts to $\sim Q \rightarrow P$ $\therefore \sim P \rightarrow Q$ (contrapositive)	valid argument

While using except, care must be taken to convert the statement to a "if- then" statement.

Example:

All the trees must be trimmed except the oak tree.
converts to
If it is not an oak tree, it must be trimmed
Therefore, if it is not trimmed, it is an oak tree (contrapositive inference)

In symbolic form:

$\sim oak \rightarrow trimmed$

$\therefore \sim trimmed \rightarrow oak$ (contrapositive inference)

Verbal Reasoning
© Gift Of Logic, Inc * Copying prohibited

CONDITIONAL ARGUMENTS — Conjunction in Antecedent

Argument Structure	Symbolic	
If P and Q then R Not R Therefore, Not P or Not Q	$P \& Q \rightarrow R$ $\therefore \sim R \rightarrow \sim P \parallel \sim Q$	valid argument

If both P and Q happen, then R must happen.
Therefore, if R does not happen, then we can infer that either P or Q or both (inclusive OR) did not happen.

Example:

If the Principal comes <u>and</u> the Mayor comes, the awards will be given. Therefore, if the awards were not given, then either the principal did not come <u>or</u> the mayor did not come or both did not come.

Note the use of "or" and "and" in the above argument.

In symbolic form:

    principal & mayor $\rightarrow$ awards given
    ~ awards given $\rightarrow$ ~ (principal & mayor) (contrapositive)
    ~ awards given $\rightarrow$ ~ principal $\parallel$ ~mayor (contrapositive)

Note how the & in the conditional has become an $\parallel$ in the contrapositive inference.

Verbal Reasoning

CONDITIONAL ARGUMENTS — Conjunction in Consequent

Argument Structure	Symbolic	
If P then Q and R	$P \rightarrow Q \ \& \ R$	valid argument
Not Q or Not R	$\therefore \sim Q \parallel \sim R \rightarrow \sim P$	
Therefore, Not P		

If P happens, then Q and R both must happen. Therefore, if either Q or R or both (inclusive OR) do not happen, then we can infer that P did not happen.

Example:
If Hillary wants to attend the ceremony, she must wear a red shirt <u>and</u> a red pant. Therefore, if Hillary did not wear a red shirt <u>or</u> a red pant or both, then she cannot attend the ceremony.

Note the use of "or" and "and" in the above argument.

In symbolic form:
  attend ceremony $\rightarrow$ red shirt & red pant
  $\sim$ (red shirt & red pant) $\rightarrow$ $\sim$attend ceremony (contrapositive)
  $\sim$ red shirt $\parallel$ $\sim$ red pant $\rightarrow$ $\sim$attend ceremony (contrapositive)

Note how the & in the conditional has become an $\parallel$ in the contrapositive inference.

Verbal Reasoning

CONDITIONAL ARGUMENTS   Disjunction in Antecedent

Argument Structure	Symbolic	
If P or Q then R Not R Therefore, Not P and Not Q	$P \parallel Q \rightarrow R$ $\therefore \sim R \rightarrow \sim P \,\&\, \sim Q$	valid argument

If P happens or Q happens, then R must happen. Therefore, if R did not happen, we can infer that both P and Q did not happen.

Clearly, since at least one of either P or Q is sufficient for R to happen, we can conclude by contrapositive inference that if R did not happen, then both P and Q did not happen.

Example:

If Moses sings <u>or</u> Mary sings, then Roy will sing. Therefore, we can infer that if Roy did not sing, then both Moses <u>and</u> Mary did not sing.

Note the use of "or" and "and" in the above argument.

In symbolic form:

$$\text{Moses} \parallel \text{Mary} \rightarrow \text{Roy}$$
$$\sim \text{Roy} \rightarrow \sim (\text{Moses} \parallel \text{Mary}) \quad \text{(contrapositive)}$$
$$\sim \text{Roy} \rightarrow \sim \text{Moses} \,\&\, \sim \text{Mary} \quad \text{(contrapositive)}$$

Note how the $\parallel$ in the conditional has become an & in the contrapositive.

Verbal Reasoning

CONDITIONAL ARGUMENTS — Disjunction in Consequent

Argument Structure	Symbolic	
If P then Q or R or both Not Q and Not R Therefore, Not P	P → Q ‖ R ∴ ~Q & ~R → ~P	valid argument

If P happens, then Q or R or both must happen. Therefore, if Q did not happen and R did not happen, then we can infer that P did not happen.

Example:

If the awards are to be given, then either the principal or the mayor must come. So, if the principal and the mayor do not come, then the awards will not be given.

Note the use of "or" and "and" in the above argument.

In symbolic form:
> awards → Principal ‖ Mayor
> ~(Principal ‖ Mayor) → ~ awards (contrapositive)
> ~Principal & ~Mayor → ~ awards (contrapositive)

Note how the ‖ in the conditional has become an & in the contrapositive.

Verbal Reasoning
© Gift Of Logic, Inc * Copying prohibited

CATEGORICAL ARGUMENTS — two categories

Arguments involving categories (groups) are called categorical arguments. Categorical arguments can have categorical statements in their premises or conclusion or both.

The relationship between two categories P and Q can be described by the following categorical statements:

    All P are Q   No P are Q
    Some P are Q   Some P are not Q

Relationships between two categories are represented using Venn Diagrams using the following conventions.

* Represent the relationship between categories using two intersecting circles. Each circle represents one category. There may or may not be members common to both categories.
* Black out the area which has no members.
* Place an "x" in the area that is represented by the "some" relationship.
* Areas that are not shaded represent the unknown. There may or may not be members in it.

Using this convention, we can represent the categorical relationships using Venn diagrams as shown next. Doing so will help us to make inferences regarding these relationships.

Verbal Reasoning

CATEGORICAL ARGUMENTS — two categories

The Venn diagram for the relationship between two categories P and Q are shown below.

All P are Q  No P are Q 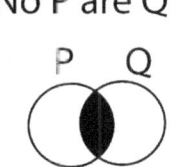 Some P are Q 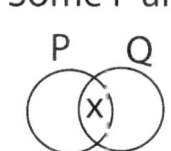 Some P are not Q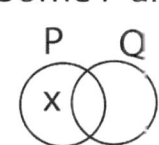

<u>All P are Q</u>: This means that, there is no P that is not a Q. The blacked out portion represents P that is not Q. Since there is no member in this area, it is blacked out.

<u>No P are Q</u>: This means that there is nothing in common between P and Q. This is represented by blacking out the intersecting portion of P and Q.

<u>Some P are Q</u>: This means that there is at least one member that belongs to both P and Q. This is represented by a "x" in the intersecting portion of P and Q.

<u>Some P are not Q</u>: This means that there is at least one member that is P, but not Q. This is represented by a "x" in the portion of the diagram that falls under P, but is outside of Q.

CATEGORICAL ARGUMENTS — two categories-inferencing

We saw how to represent categorical statements using Venn diagrams. When categorical statements are used in arguments either as premises or conclusion or both, Venn diagrams are useful in verifying whether the inferences are valid or invalid.

Following is the procedure to find out if the conclusion (inference) of a categorical argument is valid or invalid:

> Draw all the premises only in a Venn Diagram.
> After drawing the premises, do you see the conclusion?
> > If the answer is Yes, then it is a valid argument.
> > If the answer is No, then it is an invalid argument.

In other words, after drawing the premises, if the conclusion is drawn automatically, then it is a valid categorical argument. If not, it is an invalid categorical argument.

Following is the procedure to draw an inference (conclusion) from one or more categorical statements.

Draw all the premises in a Venn Diagram.

After drawing the premises, look for areas not described by the premises that are blacked out, or have an "x" in them.

These areas represent the inferences that can be drawn from the premises. Explain these areas verbally using categorical statements (all, some, no).

| CATEGORICAL ARGUMENTS | two categories-inferencing |

Consider the following categorical argument.

All P are Q.
Therefore, all Q are P.

Is this argument valid or invalid? First draw the Venn diagram for the premise All P are Q. Since this is the only premise, after drawing this premise, is the conclusion drawn? If yes, then it is a valid argument. If not, it is an invalid argument.

All P are Q

After the premise is drawn, the conclusion All Q are P is not drawn automatically. The diagram for the conclusion will look like the one shown below and it is not drawn/seen automatically after the diagram of the premise (shown above) is drawn. So, this is an invalid argument.

All Q are P

CATEGORICAL ARGUMENTS — two categories-inferencing

Some P are Q.
Therefore, some Q are P.

Verify using a Venn diagram whether this argument is valid or invalid.

Some P are not Q.
Therefore, some Q are not P.

Verify using a Venn diagram whether this argument is valid or invalid.

All P are Q. All Q are P.
Therefore, no P are Q.

Verify using a Venn diagram whether this argument is valid or invalid.

CATEGORICAL ARGUMENTS — three categories

The relationships described in categorical arguments involving three categories may be difficult to visualize. For example, given two categorical premises involving three categories, how do we derive a conclusion and know for sure that the conclusion is valid? Venn Diagrams, similar to the one drawn to represent the relationship between two categories, are helpful to draw inferences in this scenario as well.

When there are three categories P, Q, and R, all the relationships between these categories can be represented by three intersecting circles. After we represent the premises in a Venn Diagram, if we can automatically see the conclusion (without drawing anything else), then the argument is valid. If not, it is invalid.

In a situation where an "x" (representing the "some" relationship) can be placed in more than one location, evaluate the validity of the argument by placing the "x" in each possible location. For example, the "Some P are Q" relationship can be drawn by placing an "x" in the area marked as PQ or PQR. In this case, place an "x" in PQ first and evaluate the validity and then place an "x" in PQR and evaluate the validity again.

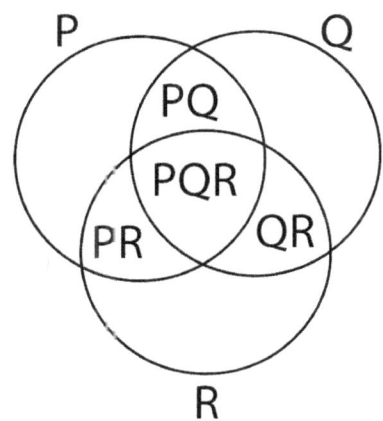

Verbal Reasoning
© Gift Of Logic, Inc * Copying prohibited

CATEGORICAL ARGUMENTS — three categories-inferencing

The three circles represent each of the three categories, P, Q, and R respectively. Areas that are common to two categories are marked as PQ, PR, and QR respectively. The area that is common to all three categories is marked as PQR. The area outside the three circles within the square represents everything that is outside of P, Q, and R (that is, not-P, not-Q, and not-R). It is important to note the location of these areas.

Verify whether the following argument is valid or invalid.
>   All P are Q.
>   No Q are R.
>   Therefore, no P are R.

Arguments of the above structure, where there are three categories, and each of the categories appear twice in the argument are called syllogisms.

The Venn diagram for this syllogism is shown in two steps. The first premise, All P are Q is drawn first by blacking out the area that represents P that is not Q. Then, the second premise, No Q are R is drawn by blacking out the are that is common to Q and R. After drawing these premises, we need to verify if the conclusion No P are R is drawn automatically. The area common to P and R is automatically blacked out. So, this conclusion is valid, and therefore the argument is valid.

Verbal Reasoning
© Gift Of Logic, Inc * Copying prohibited

CATEGORICAL ARGUMENTS — three categories-inferencing

We started off with two premises that described the relationship between P, Q, and R. By drawing the Venn diagram, we inferred the conclusion about the relationship between P and R, which was not stated in the premises.

Thus, Venn diagrams are very useful in inferencing unstated relationships involving categories.

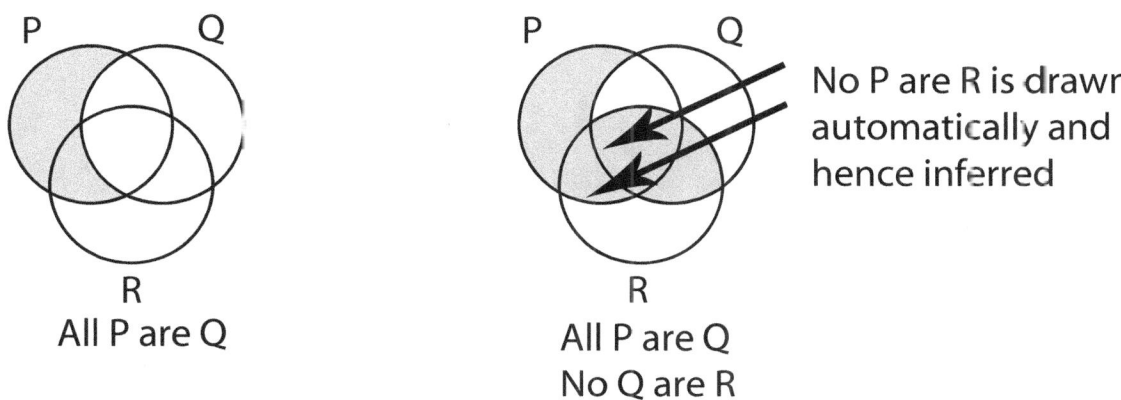

Note that when an area is not shaded, we cannot make any inference about that area. For example, in the above figure, we cannot infer that "Some R are not P", because there is no "x" in that area. We also cannot make the following inferences:
    a) "Some R are not Q", because there is no "x" in that area.
    b) "No P are Q", because this area is not fully shaded.

CATEGORICAL ARGUMENTS — three categories-inferencing

Let us consider the following problem, which presents a scenario that is similar to the argument described previously.

All players are tall.
No tall people are men.

If the above premises are correct, which of the following inferences can be drawn?
   A) Some players are men.
   B) No players are men.

The given premises and the question that is asked can be represented in symbolic as follows, where P stands for Players, T stands for Tall people, and M stands for Men:

   All P are T
   No T are M

Therefore, which of the following can be inferred?
   A) Some P are M.
   B) No P are M.

CATEGORICAL ARGUMENTS   three categories-inferencing

Draw the Venn diagram of the two premises, and see if one of the answer choices is automatically drawn. From the diagram, we cannot deduce that some P are M (choice A), because the area between P and M is shaded, which means that no Players are Men. Choice B, No P are M, is automatically drawn and therefore, it is the correct answer.

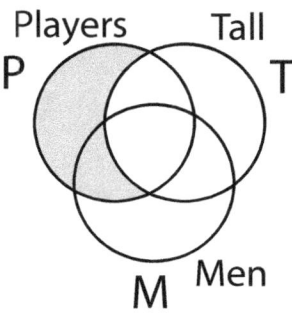

All Players are Tall
(All P are T)

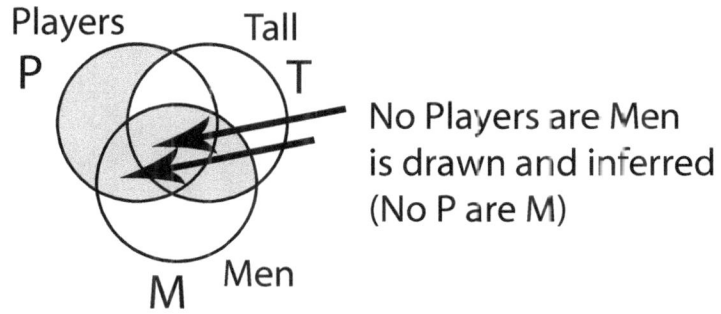

All Players are Tall (All P are T)
No Tall people are Men (No T are M)

Note that the first diagram shows the first premise, and the second diagram shows the first and the second premise. This is done to explain the concept. It is not necessary to draw two Venn diagrams. You can just draw one Venn diagram with both the premises, and make the inference using this diagram itself.

You can now practice solving the two problems in the next page.

CATEGORICAL ARGUMENTS — three categories-inferencing

Draw a Venn diagram to verify that the following categorical argument is valid.

All P are Q.
All Q are R.
Therefore, All P are R

Draw a Venn diagram to verify that the following categorical argument is invalid.

No P are Q.
No Q are R.
Therefore, no P are R.

Verbal Reasoning

OTHER DEDUCTIVE ARGUMENTS

Some arguments may not contain any conditional statements or categorical statements. They may contain just factual statements that describe the relationship between objects based on their properties like height, weight etc. In the case of these arguments, we use common sense along with basic, commonly understood principles to decide if the argument is valid or invalid.

Example: Mount Everest is the tallest mountain in the world. So, it follows that the elevation of an object that is on top of Mount Everest is more than the elevation of the same object placed on top of any other mountain.

In the above argument, there is no conditional or categorical statement involved. Using basic knowledge involving elevation, we can infer that the argument is valid.

Example: February comes after January. March comes after February. Therefore, March comes after January.

In the above argument, there is no conditional or categorical statement involved. Just using common knowledge regarding the order of months, we can infer that the argument is valid.

CAUSAL ARGUMENTS — causal statements

Causal statements describe the cause and effect relationship between two events.

Example: The hurricane caused widespread damage.

In this example, the "hurricane" is the cause and "widespread damage" is the effect. The causal statement can also be stated a bit differently as follows:

> Widespread damage was caused by the hurricane.

In the above statement too, the "hurricane" is the cause and "widespread damage" is the effect. But, it is stated in such a way that the effect appears first in the statement, and the cause appears later.

The following are some words/phrases that are typically used to indicate causality:
  causes, caused, caused by, because of, effect, responsible for

The following are examples of some more causal statements, their causes and effects.

Causal statement	Cause	Effect
Inflation causes prices to rise.	Inflation	Price rise
The tire got punctured because of the nail.	Nail	Puncture
Jay is responsible for this disaster.	Jay	Disaster

Verbal Reasoning

CAUSAL ARGUMENTS causal-inferencing

Cause and effect can be represented in symbolic as follows: c→.

Causal Statement: P causes Q
Symbolic Representation: P c→ Q

The symbol c→ is used in this book to represent causality. If P represents the cause and Q represents the effect, then the causal relationship is represented by P c→ Q. Note that the cause happens first, and the effect happens after the cause.

When several events have a cause and effect relationship among themselves, we can make certain inferences from these relationships.

1) <u>Reverse causal inference is incorrect.</u>
  If P causes Q, then we cannot infer that Q causes P.
    P c→ Q
  ∴ Q c→ P is an incorrect causal inference.

P occurred first and caused Q to occur later; not the other way around. The timing of the cause and effect is important. The cause happens first, followed by the effect.

CAUSAL ARGUMENTS — causal-inferencing

Example:

Causal: The bus accident caused a rush of patients to the hospital.
Reverse causal: The rush of patients to the hospital caused the bus accident.

Clearly, the reverse causal inference is incorrect.

2) Correlation does not imply causation.
 Just because two events P and Q have a correlation does not mean that P causes Q.

Example:
   Several students in this college wear white jeans.
   Several students in this college are extremely smart.
   Therefore, white jeans causes the students to be extremely smart.

The correlation between the two premises is that the students belong to the same college. This correlation is incorrectly assumed to be the reason why white jeans causes the students to be extremely smart.

CAUSAL ARGUMENTS — causal-inferencing

3) Successive events do not imply causation.
P happens first. Q happens next. Therefore, P is the cause of Q.
This is an incorrect inference. Just because two events happen one after the other does not mean that one causes the other.

Example: The car accident happened at 8 AM. The bus accident happened at 8:15 AM. Therefore, the car accident caused the bus accident to happen.

This conclusion cannot be inferred. Just because the two accidents happened one after the other does not mean that one caused the other.

4) Cause and Effect can be chained.
P caused Q. Q caused R. Therefore, P caused R.

This is a correct causal inference. In this case, P is the root cause of R. Q is the immediate cause of R. Note that Q is the effect in P c→ Q and the cause in Q c→ R.

Symbolic:
P c→ Q ; Q c→ R ; ∴ P c→ R is a correct causal inference. P is the "root cause" of R.

Example: Virus caused the fever. Fever caused the cough. Therefore, virus caused the cough.

This is a correct causal inference. Virus is the "root cause" of the cough.

CAUSAL ARGUMENTS — examples

Causal arguments can be described as strong or weak. They can be described in terms of probability as well. It is not possible to say with certainty that they are valid or invalid. Whether they are strong or weak depends on the strength or weakness of the premises.

In a causal argument, causal statements can appear in the premises or in the conclusion or both. When an argument has a causal conclusion, the argument can be strengthened or weakened by providing additional premises to strengthen or weaken the conclusion.

Example: Strong argument with a causal premise:

> Heavy traffic definitely causes headaches. (causal premise)
> London has heavy traffic.
> Therefore, the people of London have headaches.

This causal argument is very strong. If the premises are true, then the conclusion has a very high probability of being true. Words like "definitely" are helpful in ascertaining the strength of the argument.

CAUSAL ARGUMENTS — examples

Example: Strong argument with a causal premise:

    Cold weather will cause the price of fruits to go up.
    Weather is cold during the winter months.
    So, the price of fruits will go up during winter months.

This argument is strong. There is a high probability for the conclusion to be true if the premises are true.

Example: Weak argument with a causal conclusion:

Ten percent of the apples grown in this farm are organic.
Consuming organic fruits leads to better health.
Therefore, the apples grown in this farm are responsible for better health.

This argument is weak because it considers a small percentage as sufficient proof to make the conclusion. If ninety percent of the apples grown in this farm were organic, the argument would be strong.

CAUSAL ARGUMENTS — examples

Example: Invalid argument with causal conclusion:

Consider another causal argument with causality in its conclusion.

> The car accident happened at 8 AM.
> The bus accident happened at 8:05 AM.
> Therefore, the car accident caused the bus accident to happen.

This argument is invalid. The conclusion of this argument describes a causal relationship. The cause is the car accident and the effect is the bus accident. This conclusion is uncertain because just because these two accidents happened one after the other does not mean that one is caused by the other.

Strong arguments can be weakened by introducing additional premises.

Weak arguments can be strengthened by introducing additional premises.

You will encounter strengthen and weaken logical reasoning problems in workbook #6.

Verbal Reasoning

STATISTICAL ARGUMENTS

Statistical arguments are based on premises that refer to results of surveys, statistical figures such as maximum, minimum, average, etc and on numeric expressions such as very small, very large, tiny, small fraction, etc. They are inductive arguments because their conclusion cannot be determined as true or false. If the survey sample is increased, the conclusion will appear strong, whereas if it is decreased, the conclusion will appear weak. Following are some examples of statistical arguments.

Example
1) Ninety seven percent of people who live in Columbia Road own two cars. Jenny lives in Columbia road. So, Jenny owns two cars.

Since a majority of people living in Columbia Road own two cars, it is very likely that Jenny, who also lives in Columbia Road, also owns two cars. This conclusion is strong.

2) Only five percent of the students in a class have dancing skills. John is a student in this class. So, John also has dancing skills.

Since most of the students do not have dancing skills, John is also not likely to have dancing skills. So, this conclusion is weak.

ANALOGOUS ARGUMENTS

Analogous arguments base their conclusions on the analogies between the subjects described in the premises. They are inductive arguments whose conclusion cannot be determined to be true or false.

These arguments can be described using terms such as strong, weak, likely, highly likely, unlikely, highly unlikely etc.

Example of an analogical argument:
>Hotel A is a five star hotel.
>Hotel B is also a five star hotel.
>Hotel A is popular among tourists.
>Therefore, hotel B will also be popular among tourists.

The analogy is that the two hotels are five star hotels. After the similarity is established, a new fact about hotel A is described- that it is popular with tourists. The analogy is used as the reason to make the conclusion that hotel B will also be popular among tourists.

In arguments like this, we cannot be certain about the conclusion. We can strengthen or weaken the conclusion. We can strengthen the argument by introducing a new fact that hotel B is in the same neighborhood as hotel A. We can weaken the argument by introducing a new fact that hotel B does not have a swimming pool.

Verbal Reasoning

ANALOGOUS ARGUMENTS

Example:
>Jack and James are both men of the same race.
>Jack is tall.
>Therefore, James is tall.

In this argument, the similarity between Jack and James is that they are both men of the same race. This similarity is used to conclude that if one is tall, the other one will also be tall. The relevance of the similarity to the conclusion is weak, because we know that in any race there are tall men and short men. This is a weak analogical argument.

Example:
>Jerry has sung one hundred songs.
>Tom has also sung one hundred songs.
>Jerry is a famous singer.
>Therefore, Tom is also a famous singer.

This argument bases its conclusion on the fact that both Tom and Jerry have sung one hundred songs. We cannot conclude for sure whether Tom is also a famous singer from the facts. It can be strengthened or weakened by providing additional information.

Thus, it can be understood that analogous arguments have to be assessed based on the strength of their analogy.

FLAWED ARGUMENTS

Flaws occur in arguments due to various reasons. Sometimes, the flaws occur because of incorrect reasoning. Sometimes, flaws are introduced deliberately to deceive or convince. So, it is important to learn to identify the flaws in arguments.

Consider the following example:

   John is a liar. So, he is not in our soccer team.

We know from the material covered so far that a valid argument is one where, if the premises are true, the conclusion is also true. Applying this rule to the above argument, we cannot infer that if John is a liar then he will not be in the soccer team. This argument is invalid. The reasoning is flawed because it does not provide sufficient evidence to reach its conclusion.

In this section we refer to invalid arguments as "flawed arguments". We will learn to find out why an argument is flawed and in the process learn about various types of flaws that can be found in arguments. Flaws are also referred to as fallacies.

FLAWED ARGUMENTS

Circular Argument Fallacy (begging the question fallacy)

The reasoning in a <u>circular argument</u> is flawed because the premise states the conclusion and the conclusion states the premise.

Structure:
Premise states that the conclusion is true. The conclusion states that the premise is true.

Example:
I refuse to pay the tax. Therefore, I will not pay the tax.

In this argument, the premise (I refuse to pay the tax) explicitly states the conclusion (I will not pay the tax). The conclusion also states the premise.

Example:
　　Sanchez is the most talented player. So, he is also the best player.

In this argument, the premise (Sanchez is the most talented player) explicitly states the conclusion (He is also the best player). The conclusion also restates the premise.

FLAWED ARGUMENTS

Appeal to emotion fallacy

The reasoning in an <u>appeal to emotion</u> argument is flawed because it depends on emotions instead of facts.

Structure:
Premise states that something is desired. Conclusion states that because something is desired, it must be good.

Example:
The doors of this new car model open upwards. If you buy this car, everyone will admire you for being fashionable. So, you must buy it now.

Note that the reason suggested for selling the car appeals to emotion and does not talk about the merits of the car.

FLAWED ARGUMENTS

Appeal to flattery fallacy

The reasoning in an <u>appeal to flattery</u> argument is flawed because flattery is not sufficient evidence to prove the conclusion.

Structure:
In the premise, one person flatters another person. In the conclusion, a favorable request is made.

Example:
Emily, you are a great interior designer. Your house is beautiful. The landscape is breathtaking. So, you must donate money to my college.

In this flawed argument, the person making the argument flatters Emily and uses this flattery to request money. Flattery is not evidence to justify the conclusion.

FLAWED ARGUMENTS

Appeal to authority fallacy

The reasoning in an <u>appeal to authority</u> argument is flawed because the conclusion of the argument refers to an expert whose expertise is not relevant to the argument.

Structure:
Premise states that a person is an authority in some subject. Conclusion states that his advice is therefore correct in a totally different subject.

Example:
Dr. Jacob says that this building is unsafe. Since he is an expert cardiologist, we must accept his opinion about the safety of this building.

In this argument, the advice of an expert (cardiologist), whose expertise is irrelevant in the area of building safety is used to state the conclusion.

FLAWED ARGUMENTS

Straw man fallacy (besides the point fallacy)

The reasoning in a <u>straw man</u> argument is flawed because one person's opinion is deliberately misinterpreted in order to attack that person's opinion.

Structure:
Premise stated by one person is deliberately misinterpreted and a conclusion is stated based on this misinterpretation.

Example:
Our soccer coach says that we should practice for several hours everyday. But, practicing almost all day everyday is very difficult. So, we must not follow his advice.

In this argument, the coach wants everyone to practice several hours every day. But, the person making this argument does not like this. So, he distorts it by saying that the coach wants them to practice almost all day everyday. "Several hours everyday" does not necessary mean "almost all day everyday", but this straw man is used to make the conclusion.

FLAWED ARGUMENTS

Appeal to crowd fallacy

The reasoning in an <u>appeal to crowd</u> argument is flawed because it appeals to the opinion of the crowd instead of providing evidence for the conclusion.

Structure:

Premise states that most people like "something". Conclusion states that "something" must be true.

Example:

All my friends think ice cream is good for health. Therefore, I also think that ice cream is good for health.

This argument is flawed because it appeals to the crowd's opinion (all my friends) to justify its conclusion.

FLAWED ARGUMENTS

Composition fallacy

A <u>fallacy of composition</u> occurs when a conclusion is made on the assumption that if something is true for the parts, then it is true for the whole as well.

Structure:
The premise states that each part of a whole has a certain feature. The conclusion states that the whole also has this feature.

Example:
Every room in this house is well designed. So, the entire house is well designed.

Just because every room is well designed does not mean that the entire house is well designed. It is possible that the rooms are not conveniently located, and thus the design of the house may be flawed even though each room is well designed.

This argument has a fallacy of composition. It assumes that the characteristics of the parts (rooms are well designed) are also true for the whole (house is well designed).

FLAWED ARGUMENTS

Division fallacy

A <u>fallacy of division</u> occurs when a conclusion is made on the assumption that if something is true for the whole, then it also true for its parts.

<u>Structure:</u>
The premise states that the whole has certain features. The conclusion states that the parts of the whole also have the same features.

<u>Example:</u>

    My car is an old car. Therefore, every part in my car is also old.

The car may be old, but that does not mean that every part in it is also old. For example, the car could have new tires. This argument contains the fallacy of division. It assumes that what is true for the whole is also true for its parts.

<u>Example:</u>

    Mario is strong. So, every part of his body is also strong.

Mario may be strong overall, but he may have a weak eyesight. So, we cannot conclude that if he is strong, then every part of his body will be strong.

FLAWED ARGUMENTS

Attack on character fallacy (Ad-Hominem fallacy)

Arguments that attack the character of a person to justify their conclusion are called <u>Ad-Hominem</u> arguments. They are flawed because attacking a person is not a valid reasoning strategy.

Structure:
A premise states the opinion of one person. Another premise attacks this person's character and concludes that what this person says is false.

Example:
  Mary: We should have longer sports classes because sports is good for health.
  Jane: You want longer sports classes because you are a liar. So, we should not have longer sports classes.

Example:
Steve wants to sell tables very cheaply. But, Steve is a thief and the tables are also very likely stolen from someone. Therefore, we should not buy the tables sold by Steve.

In the above examples, the character of a person is attacked to prove the conclusion and hence these are flawed arguments.

FLAWED ARGUMENTS

Changing the subject fallacy (Red Herring fallacy)

Arguments that have the <u>changing the subject</u> fallacy present information with the intention of diverting the focus from the main argument. This strategy results in the original argument being forgotten. This fallacy is also called the red herring or wild goose chase fallacy.

Structure:
A premise presents an opinion. Another premise changes the subject by introducing another opinion and makes a conclusion based on this new opinion.

Example:
    Ken: Red apples are more delicious than green apples.
    Paul: Things that are red in color are very attractive. That is why I painted my car red. So, red is definitely a prestigious color.

In this argument, Ken's subject of discussion, "red apples are more delicious than green apples" was changed by Paul to "things that are red are very attractive." He then concludes that "Red is a prestigious color". Paul commits a "changing the subject" fallacy in his argument by changing the subject that is being discussed.

Verbal Reasoning

FLAWED ARGUMENTS

Hasty Generalization fallacy

An argument has a <u>fallacy of hasty generalization</u> when it makes a conclusion regarding a group based on a small sample of that group.

Structure:
A small sample is taken from a group and the results of observation from this small sample is applied on the entire group. This is flawed reasoning because the results from a small sample is hastily generalized on a larger group.

Example:
Ten percent of men surveyed in USA like to grow a beard. So, all the men in USA like to grow a beard.

This argument is flawed because the result of a survey on just ten percent of the men in USA was taken and hastily generalized (applied) on all the men in USA.

FLAWED ARGUMENTS

Faulty Analogy fallacy

<u>Faulty analogy</u> is a fallacy that results when the relevance of similarities between the analogies is not strong.

<u>Example:</u>
Susan is a baby girl. So, she loves to drink milk.
Mary is a college girl. So, she also loves to drink milk.

In this argument, Susan is a baby girl and Mary is a college girl. Even though both of them are girls, this similarity is not strong enough to prove that since Susan loves to drink milk, Mary also loves to drink milk.

<u>Example:</u>
Mike is a smart Engineer. John is also a smart Engineer. That is why John would not have stolen the money.

This argument is flawed because the similarity of the analogy, that both are smart Engineers is not strong enough to prove that John would not have stolen the money.

FLAWED ARGUMENTS

Causal fallacy

Causal fallacy occurs in an argument when two events happen at the same time and the argument concludes that one event caused the other to happen.

Causal fallacy also occurs when two events happen one after the other and the argument concludes that the earlier event is the cause of the latter event.

Example:
   The power went off at 9 AM.
   The fire started at 9:05 AM.
   Therefore, the power outage caused the fire.

In this argument, clearly there is causal fallacy. Just because two incidents occur one after the other does not imply that one incident caused the other to occur.

Causal fallacies can occur when we improperly make inferences from causal relationships. These fallacies were discussed in the section on Causal Arguments earlier in this book.

FLAWED ARGUMENTS

Affirming the Consequent fallacy

<u>Affirming the Consequent</u> fallacy occurs in conditional arguments that affirm the Consequent (Q).

<u>Structure:</u>   If P then Q. Q is true. Therefore, P is true.
<u>Example:</u>   If your yard has sprinklers, it will be green.
Your yard is green. Therefore, your yard has sprinklers.

This is flawed reasoning, because the converse of a conditional statement is false. It is possible to have a green yard without having sprinklers.

Denying the antecedent fallacy

<u>Denying the Antecedent</u> fallacy occurs in conditional arguments when the argument denies the antecedent (P).

<u>Structure:</u>   If P then Q.
P is false. Therefore, Q is false.
<u>Example:</u>
If your yard has sprinklers, it will be green.
Your yard does not have sprinklers. Therefore, it is not green.

This is flawed reasoning. because the negation of a conditional statement is false. It is possible to have a green yard without having sprinklers.

Verbal Reasoning

FLAWED ARGUMENTS

Joining the bandwagon fallacy

The reasoning in a joining the bandwagon argument is flawed because, the conclusion is made because of fear of rejection by a group of peers.

Structure:

A premise states an opinion. Another premise states the opinion of others (peers) that are not in agreement with the original premise. A conclusion is then made to satisfy the peers.

Example:

The Mayor of the City of Rooster decided not to have fireworks this year. But, the mayors of nearby cities were strongly in favor of having fireworks. So, the mayor of the City of Rooster decided to have fireworks.

In this argument, due to the fear of rejection by his peers, the Mayor of the City of Rooster decided to have fireworks. This reasoning is flawed and is called the "joining the bandwagon" fallacy. The mayor joins the bandwagon of his peers because of fear of rejection by them.

VERBAL REASONING - FURTHER READING

So far, we have discussed argument structure and the different types of arguments that you would encounter in daily life. Argumentation is a fascinating subject and you can go on to get a Ph.D in this subject. However, in the meantime, to whet your appetite to gain a wider understanding of the subject of reasoning, you can do the exercises in workbooks 6,7,8,9 and 10. Each of these workbooks have a section on Verbal Reasoning that contain several exercises in argumentation that will help you to develop strong reasoning skills. The following pages give you a brief overview of the verbal reasoning sections in workbooks 6,7,8,9 and 10.

Workbook# 6

<u>Main point</u> - You develop the ability to identify the main point of an argument. An argument is given and you are asked to pick a statement that correctly rephrases the conclusion of the argument.

<u>Must be true</u> - You develop the ability to make inferences from given premises. Several premises are given and you are asked to identify the conclusion that must be true if the premises are true.

<u>Cannot be true</u> - You develop the ability to identify the incorrect inference. Several premises are given and you are asked to identify the conclusion that cannot be true, if the premises are true.

VERBAL REASONING - FURTHER READING

Workbook# 7

<u>Strengthen</u> - You develop the ability to strengthen an argument. An argument is given. You have to identify a premise that strengthens the argument the most.

<u>Weaken</u> - You develop the ability to weaken an argument. An argument is given. You have to identify a premise that weakens the argument the most.

Workbook# 8

<u>Controversy</u> - You develop the ability to identify the controversy/point at issue. Two sets of arguments are given. You need to identify the controversy between the two arguments.

<u>Paradox</u> - You develop the ability to resolve a paradox. A paradox contains a set of facts that are in contradiction to each other. You have to find a statement that would explain/reconcile the paradox.

VERBAL REASONING - FURTHER READING

Workbook# 9

<u>Assumptions</u> - You develop the ability to identify the assumption made by an argument. An argument is given, but it makes some assumption and leaps to its conclusion. You have to identify the correct assumption from a list of assumptions.

<u>Reasoning strategy</u> - You develop the ability to identify the reasoning strategy in the argument. An argument is given. You have to identify what reasoning strategy is used in the argument.

Workbook# 10

<u>Flawed arguments</u> - You develop the ability to identify flaws in arguments. An argument is given that has a flaw in its reasoning. You have to identify the correct type of flaw in the argument.

<u>Analogous arguments</u> - You develop the ability to identify analogous arguments. An argument is given. You have to identify another argument that is analogous (parallel) to this argument.

As you can see, all these workbooks have problems that sharpen your understanding of argumentation. They give you a very valuable opportunity to enrich your critical thinking and logical reasoning abilities.

ANALYTICAL REASONING

ANALYTICAL REASONING

The analytical reasoning section in workbooks 6,7,8,9 and 10 present problems in positioning and grouping. These problems require an understanding of logical operators, logical reasoning, and common sense. Solving these problems help you develop an analytic, resourceful mind. This section will provide you information on how to approach these problems.

In "positioning problems" you will be presented with a scenario where the objective will be to place a certain number of objects (people, birds, boxes etc) in certain number of positions. There may be as many positions as there are objects, so that every object gets a position. There may be more objects than available positions, so that some objects may be left without a position. There may be more number of positions than there are objects, in which case some of the positions may be left vacant.

In addition to these scenarios, you will be presented with rules regarding which object can be placed in which position. For example, a rule may state "John must sit to the left of Joe" or "Judy must not sit immediately next to Frank". You will need to consider all the rules and apply them to the objects and find the correct positions for these objects.

After the scenario and rules are presented, you will be asked questions such as "Who can sit to the right of Judy?", "If John sits to the right of Judy, then can Joe sit in the third position?" and so on.

Solving "positioning problems" will help you to tune your brain to think in a flexible manner regarding resource allocation and working with constraints.

ANALYTICAL REASONING - POSITIONING PROBLEM

<u>Symbolic Diagrams:</u> While solving positioning problems, it is very convenient to represent the objects/rules in symbolic form so that we can make inferences and solve the problem quickly. We have already seen several symbols - symbols for "and", "or", "not", conditional, biconditional etc. These symbols and the symbols associated with conditional reasoning (converse/inverse/contrapositive/contradiction) are very helpful in diagramming the scenarios present in positioning problems.

The following table shows how to represent the typical scenarios that you will encounter in positioning problems.

Not	~
And	& ,
Inclusive OR	∥
Exclusive OR	#
Conditional	→
Biconditional	↔

Analytical Reasoning

ANALYTICAL REASONING - POSITIONING PROBLEM

Judy, Kate, and Larry are to be seated in three positions.	J, K, L Use the first letter of the objects to refer to them. If there is a conflict, use different letters to identify the objects. For example, T, Y for Tony and Tommy respectively.
Judy must sit in the third chair.	J3
Judy must sit in the third chair and Kate must sit in the fourth chair.	J3 & K4 J3, K4
Judy must sit to the left of Kate.	J-K Note that the dash (-) means that there may be others between J and K. So, J,L,K is valid.
Judy must sit immediately to the left of Kate.	JK Note that there is no "-" between J and K. No one can sit between J and K. So, J,L,K is invalid.
* Kate must sit next to Judy. * Kate and Judy must sit in consecutive positions. * Kate and Judy must sit next to each other	KJ ╫ JK Note that the rule just says that Kate must sit next to Judy. It does not say whether Kate must sit to the right or left of Judy. So KJ or JK are both possible.
Judy must sit in the first chair or Kate must sit in the third chair.	J1 ∥ K3 - both are possible (inclusive or). If the condition said "but not both", the symbol would be J1 ╫ K3.

Analytical Reasoning
© Gift Of Logic, Inc * Copying prohibited

ANALYTICAL REASONING - POSITIONING PROBLEM

Kate and Judy must not sit next to each other.	~KJ, ~JK
Judy must not sit in the third chair.	~J3 Note the symbol ~ for negation.
Judy can sit in the first chair or the third chair	J1 ⊕ J3 (note the use of exclusive OR. J1 or J3 is valid, but not both.
Judy must sit in the first chair or Kate must sit in the third chair.	J1 ∥ K3 (note the symbol of inclusive OR. J1 or K3 or both are the valid)
Judy must not sit in the third chair and Kate must not sit in the fourth chair.	~J3 & ~K4 ~J3, ~K4
If Kate sits in the first chair, Frank must sit in the third chair.	K1 → F3 Incorrect to note this as K1,F3. Refer the conditional reasoning section.
Only if Kate sits in the first chair will Judy sit in the third chair.	J3 → K1 Refer the conditional reasoning section. K1 → J3 is incorrect notation.
If Kate sits in the first chair, Frank must not sit in the third chair.	K1 → ~F3
If and only if Judy sits in the first chair, Kate will sit in the third chair.	J1 ↔ K3, note the biconditional. J1→K3, K3→J1
Spot# 3 must be vacant.	X3 (X represents vacancy)
Spot# 3 must not be left vacant.	~X3, X represents vacancy.

SAMPLE POSITIONING PROBLEM

SCENARIO

Rudy, Puppy, Tommy and Tony are four dogs that need to be seated in four consecutive spots, numbered 1,2,3, and 4. Rudy must always sit in the third spot. Tommy and Tony must always sit in consecutive positions.

QUESTIONS

1) In which of the following positions can Puppy be seated?

    A) first   B) second   C) third D) fourth

2) Which of the following is always true?
    A)   Puppy always sits in the fourth position.
    B)   Tony always sits in the first position.
    C)   Tommy always sits in the second position.

Analytical Reasoning

SOLUTION TO SAMPLE POSITIONING PROBLEM

When solving analytic problems like this, we follow a three step process.
1) Represent the scenario using symbolic diagrams.
2) Assign objects to positions based on the rules.
3) Answer the questions.

We start off by using short names for the dogs - R, P, T and Y -- since Tommy and Tony both start with T, to avoid confusion, we use T for Tommy and Y for Tony. Then we represent the rules in symbolic form.

"Rudy must always sit in the third spot" is represented as R3. Tommy and Tony must always sit in consecutive positions is represented as TY ⫫ YT. We use the symbol ⫫ for exclusive OR here because only one of the possibilities is meaningful in this problem.

We then write the positions and their occupants in a table and write all the rules on the left of the table. We then assign objects to positions to the extent possible. We place R (Rudy) in the third position to satisfy the rule R3.

R,P,T,Y
R3
TY ⫫ YT

1	2	3	4
		R	

Analytical Reasoning
© Gift Of Logic, Inc * Copying prohibited

SOLUTION TO SAMPLE POSITIONING PROBLEM

Now we are ready to answer the questions.

1) In which of the following positions can Puppy be seated?
   A) first B) second C) third D) fourth

Even though there is no rule that says where Puppy can sit, it cannot sit anywhere it wants. It can only sit in a position that does not violate any of the rules. If Puppy sits in the first spot, then we have the following diagram.

R,P,T,Y
R3
TY ∦ YT

1	2	3	4
P		R	

This is not a valid diagram because if Puppy sits in the first position, then Tommy and Tony cannot sit together. Visually, try to place TY or YT - since this is not possible, Puppy cannot sit in the first position. You can thus see the power of solving this problem using symbols. Now, you would have realized that two consecutive spots need to be made available for T and Y in positions 1 and 2 since position 3 is reserved for R. So, P (puppy) can sit in the fourth spot only.

SOLUTION TO SAMPLE POSITIONING PROBLEM

Now, let us answer the next question.

2) Which of the following is always true?
   A) Tony always sits in the first position.
   B) Tommy always sits in the second position.
   C) Puppy always sits in the fourth position.

Refer back to the scenario diagram and eliminate the choices until you find the correct one.

R,P,T,Y
R3
TY ∦ YT

1	2	3	4
Y	T	R	P
T	Y	R	P

Answer choice A is wrong because Y (Tony) can sit in the first position or the second position.

Choice B is also wrong because T (Tommy) can sit in the first or the second position. The rule TY ∦ YT allows them to switch positions. Choice C is the correct answer. Puppy always must sit in the fourth position because position# 3 is taken by R and TY ∦ YT means that positions 1 and 2 are not available for P.

Workbooks 6-10 contain positioning problems that involve various types of constraints and situations such as those involving vacant positions. Solving these problems will help you to develop strong analytic skills.

Analytical Reasoning

ANALYTICAL REASONING - GROUPING/SELECTION PROBLEM

In grouping/selection problems, the scenario is such that you are given a list of objects and a set of rules that must be followed to select groups out of these objects. Then, you will need to answer questions that will ask you to verify if a group is valid or not. A variation of this problem is when you first select a group from a list of available objects and then position them in a certain number of spots.

In grouping/selection problems, the order of selection is not important. For example, the rule "Judy and Kate must be selected together" can be represented by JK. In the context of the grouping problem, it is not necessary to say KJ because the order of selection is not important, but is understood. Some of the typical rules that you encounter in grouping problems and their symbolic representation is shown in the table below.

Judy (J) and Kate (K) must be selected. Judy and Kate must be selected together.	JK; note that selecting J only without K, or selecting K only without J, is invalid. Both J and K must be selected.
Judy and Kate must not be selected.	~JK, ~J, ~K; even just one of them cannot be selected. JKL is invalid, JL is invalid, KL is invalid.
Judy and Kate must not be selected together, but can be selected separately.	~JK; J ╫ K JKL is invalid, JL is valid, KL is valid. One of them can be selected, but not both.

Analytical Reasoning

ANALYTICAL REASONING - GROUPING/SELECTION PROBLEM

If Judy is selected, Kate must be selected.	J → ~K; this means JK is valid, K without J is valid, but J alone is not valid. Refer to the conditional reasoning section.
If Judy is selected, Kate must not be selected.	J → ~K ; K → ~J; (contrapositive) this means that J can be selected without K, K can be selected without J, but J and K together cannot be selected.
Either Judy or Kate must be selected. (inclusive OR)	J | K ; note that J or K or both can be selected. J is valid, K is valid, JK is valid
Either Judy or Kate must be selected, but not both. (exclusive OR)	J ╫ K; note that J or K can be selected, but not both. J is valid, K is valid, JK is invalid

Workbooks 6-10 contain grouping problems that involve various types of constraints and situations such as those involving a combination of grouping and positioning problems. Solving these problems will help you to develop strong analytic skills. A sample grouping problem is shown next.

SAMPLE GROUPING/SELECTION PROBLEM

SCENARIO

Two people are to be selected from a group of four people named A,B,C, and D. If A is selected, then C must not be selected. If B is selected then D must not be selected

QUESTIONS

1) Which of the following choices represent a valid selection?
    A) A,B   B) A,C   C) D,B   D) C,D

2) In how many teams can A be present?
    A) 1   B) 2   C) 3

SOLUTION TO SAMPLE GROUPING/SELECTION PROBLEM

We begin by noting the objects involved and the rules that govern the selection.

Two people are to be selected from a group of four people named A, B, C, and D.

If A is selected, then C must not be selected
   >> This rule is represented as A → ~C
If B is selected then D must not be selected
   >> This rule is represented as B → ~D

1) Which of the following choices represent a valid selection?
   A) A,B   B) A,C   C) D,B   D) C,D

Answer choices B and C are clearly incorrect, since they violate the rules A → ~C and B → ~D respectively. Answer choices A and D are correct. A and B form a valid team and C and D also form a valid team, as they do not violate the rules.

2) In how many teams can A be present?
   A) 1   B) 2   C) 3

List the groups where A can be present and eliminate the ones that violate the rules.

AB - valid   AC - invalid (violates rule A → ~C)   AD - valid.
So, A can be present in two teams.

Analytical Reasoning
© Gift Of Logic, Inc * Copying prohibited

PICTORIAL REASONING

PROBLEM TYPE DESCRIPTION

In all the books in this series, there is a section on Pictorial Reasoning. The purpose of providing this section is to train your brain in non-verbal reasoning. When we look at a set of pictures, we can observe patterns, similarities, differences, oddities, and so on. These observational activities alert our mind and improves our cognitive abilities in non-verbal reasoning. The following table shows the different types of problems that you will encounter in the Pictorial Reasoning sections of Workbooks numbered 0-10. Following this table are some actual problems that you will encounter in these workbooks that you can get a feel for.

Problem Type	Problem Description	Workbook
Connect the dots	Connect the dots according to given instructions to form a figure.	0
Maze	Find your way through the maze.	0
Picture sequence	Find the next logical picture in the sequence from a set of pictures.	0,1,2,3,4,5
Picture difference	Find the difference between two pictures.	0,1,2,3,4,5
Picture analogy	Find the relation between two pictures and identify the same relation in a different set of pictures.	2,3,4,5
Odd picture	Find the odd picture from a set of pictures.	2,3,4,5
Pattern matching	Find the pattern that will logically fit into an existing pattern.	2,3,4,5

PROBLEM TYPE DESCRIPTION

Problem Type	Problem Description	Workbook
Pattern Perception missing pattern continuing pattern	Find the picture with a pattern that will match an existing pattern. Find the next pattern that will continue the pattern found in a given set of pictures.	6,7,8,9,10
Figure formation	Find the figure that will be formed by joining a set of figures.	6,7,8,9,10
Paper folding and cutting	Find the final form of a paper that is folded and then has holes cut in it.	6,7,8,9,10
Figure Matrix - Similarity	Find a picture that has the same characteristic as three other pictures.	6,7,8,9,10
Figure Matrix - Analogy	Find a picture that will complete the analogy that exists in the picture set.	6,7,8,9,10
Rule Detection	Understand a given rule and then identify the correct set of pictures that follow this rule.	6,7,8,9,10

Pictorial Reasoning
© Gift Of Logic, Inc * Copying prohibited

CONNECT THE DOTS - JUMP

Connect the dots shown below, starting from the lowest number to the next number until you reach the highest number. Stop when you reach the highest number. The letter J is an instruction to jump to the number shown next to it without drawing a line to it. For example, 2-J-3 means, jump from 2 to 3 without drawing a line from 2 to 3. After jumping, continue to the next higher number that has not been connected.

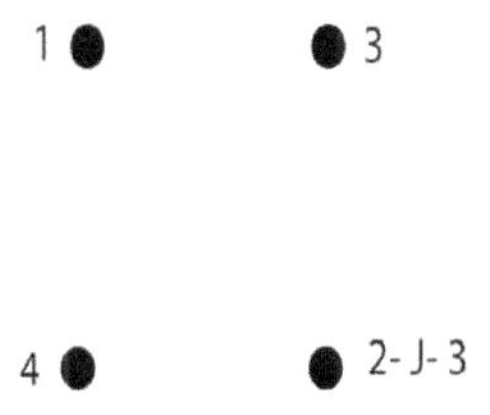

1●   ●3

4●   ●2-J-3

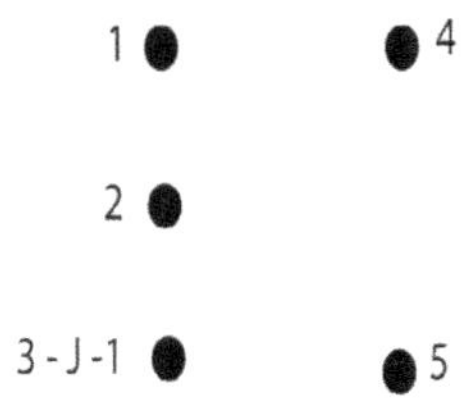

1●   ●4

2●

3-J-1●   ●5

Pictorial Reasoning
© Gift Of Logic, Inc * Copying prohibited

MAZE

Solve the mazes shown below from Start to End.

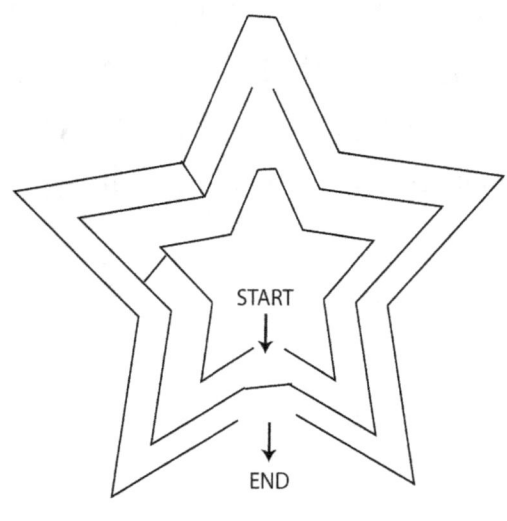

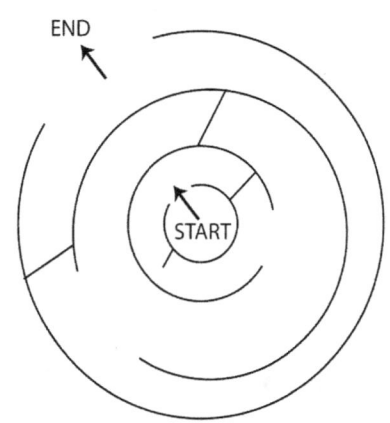

PICTURE SEQUENCE

Figure out the logic in the sequence of pictures shown, and draw the next picture in the sequence that will continue the logic.

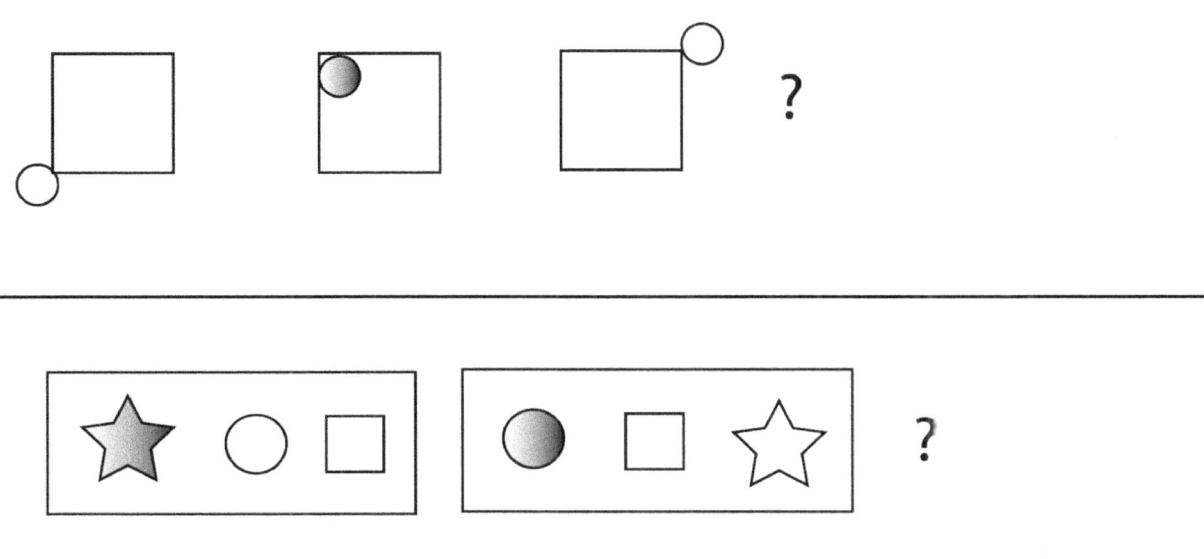

Pictorial Reasoning
© Gift Of Logic, Inc * Copying prohibited

ODD PICTURE

Circle the odd picture in each set.

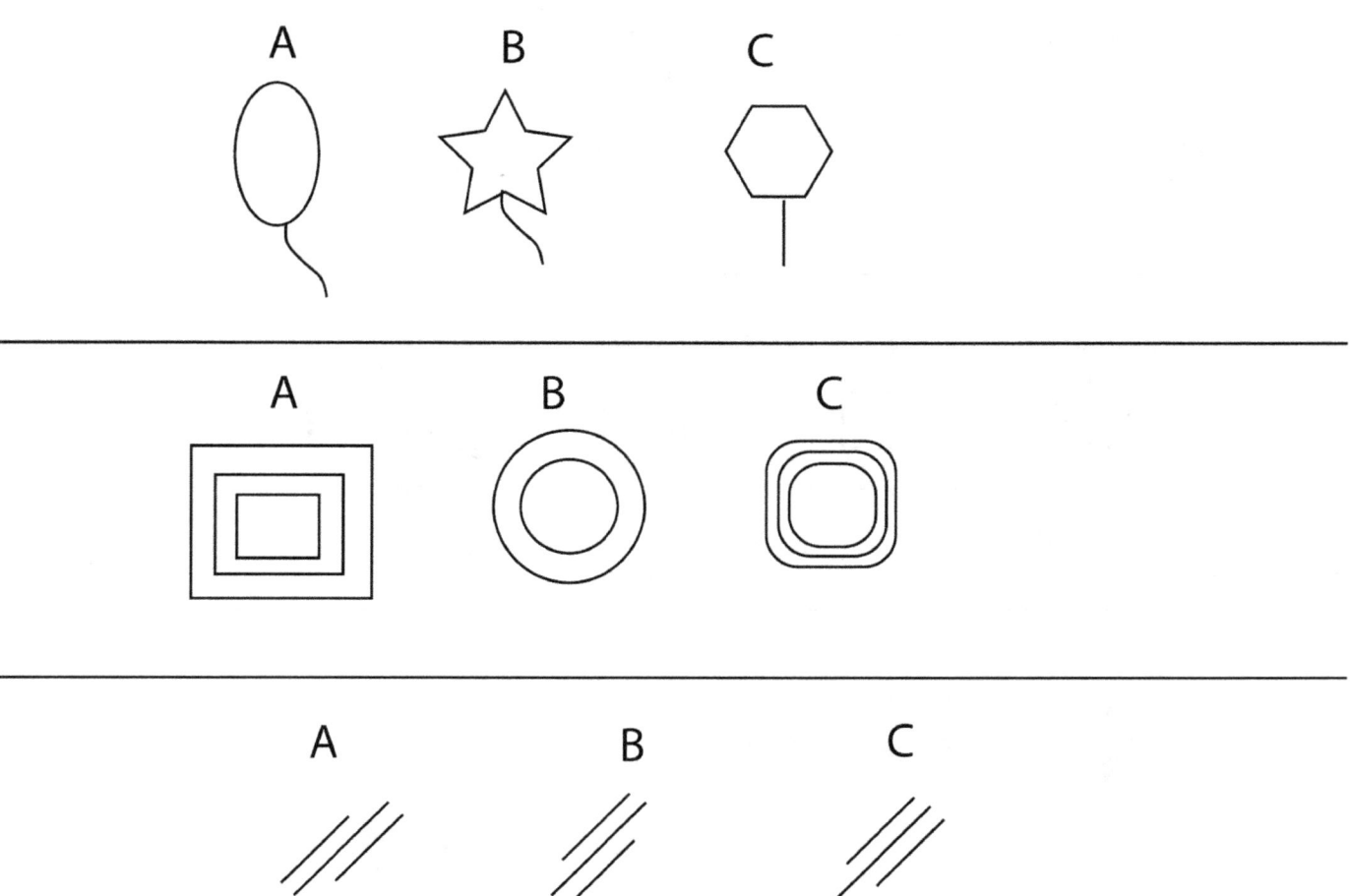

Pictorial Reasoning

PICTURE DIFFERENCE

Mark the differences between the two pictures in each set, with arrows.

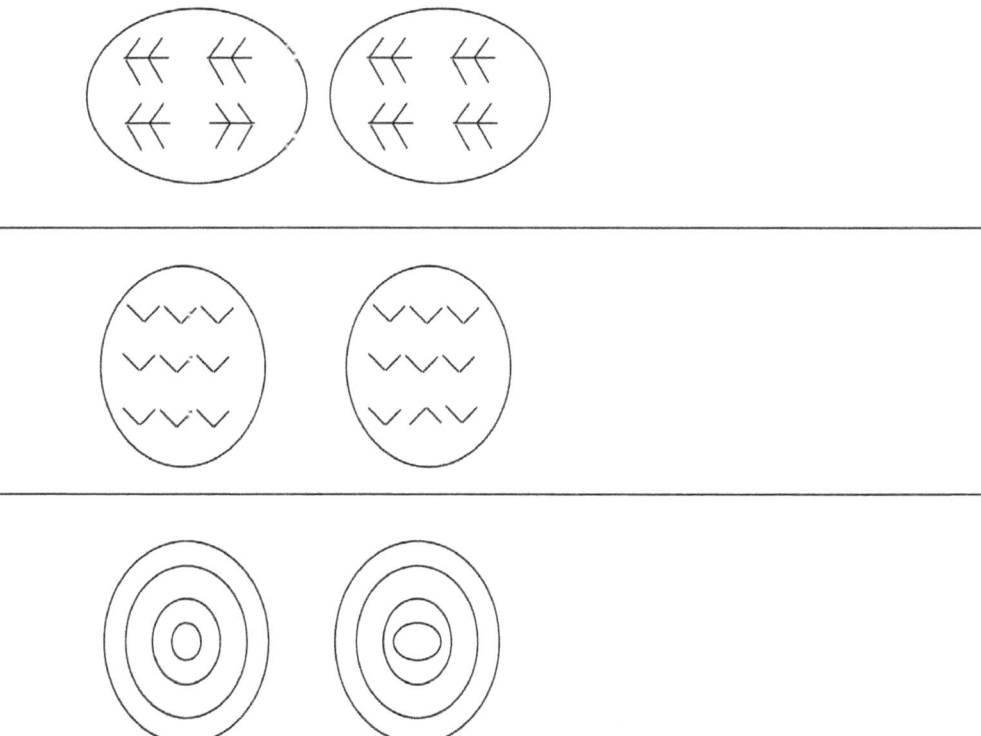

Pictorial Reasoning

PICTURE ANALOGY

Circle the correct choice (A or B) that will complete the picture analogy.

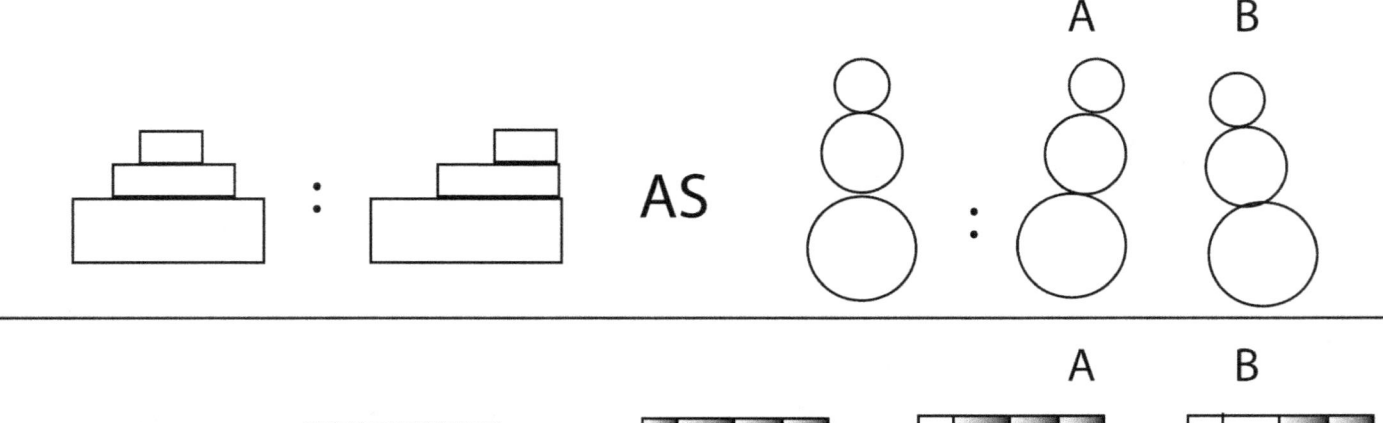

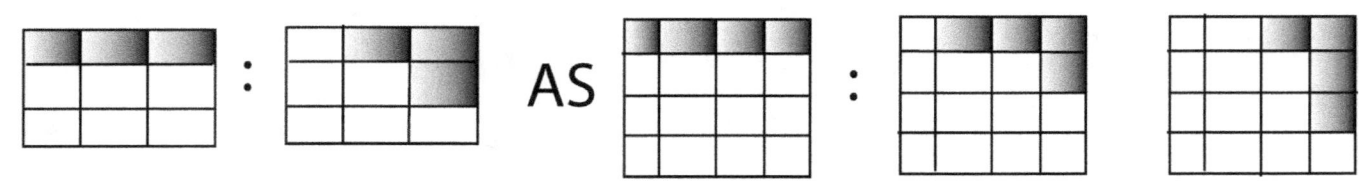

PATTERN MATCHING

Find the logical pattern in the pictures on the left, and identify the picture on the right that will fit in the space marked with ? to complete the pattern.

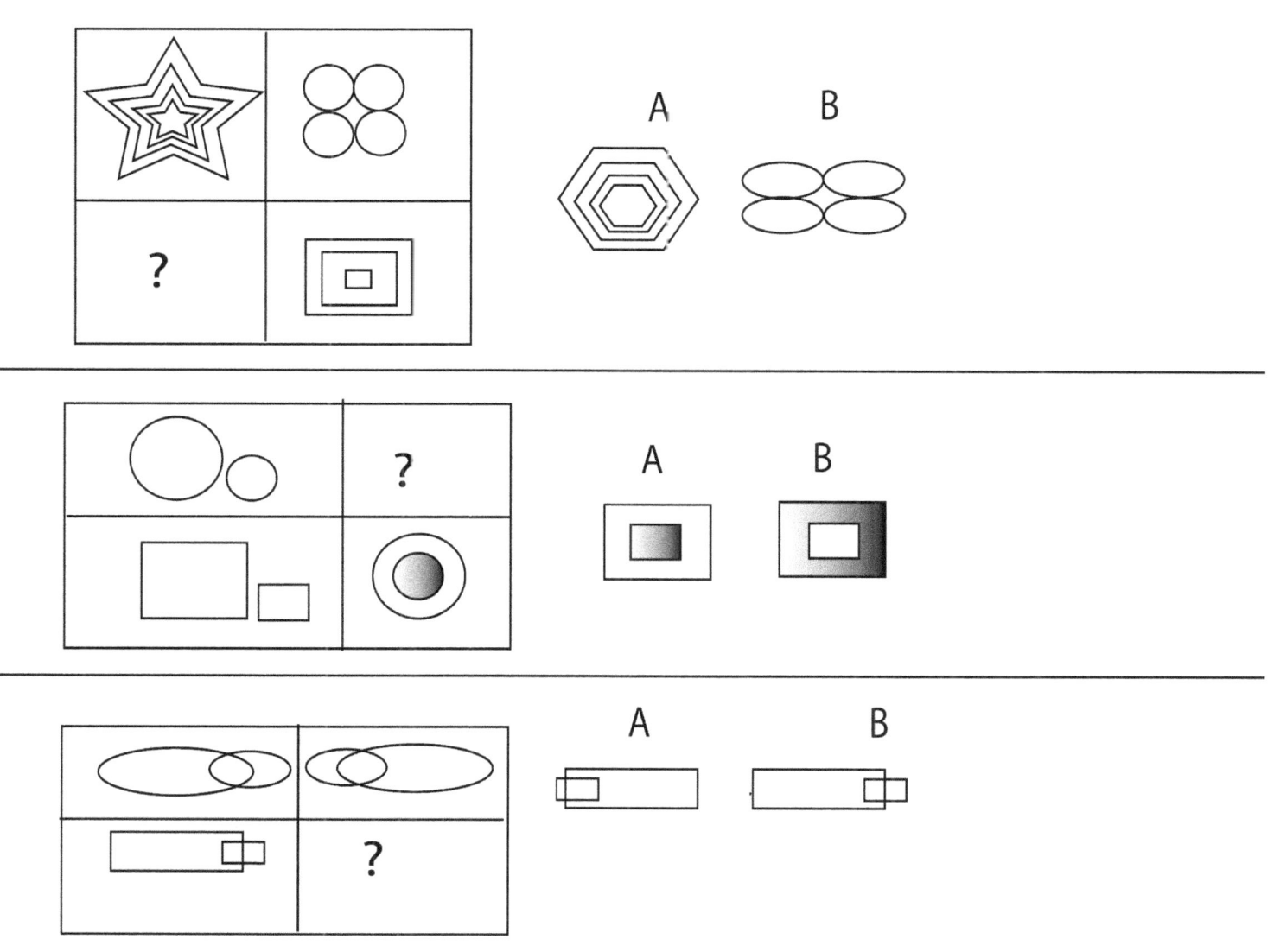

PATTERN PERCEPTION - MISSING PATTERN

Find the correct figure from the three alternatives given, that will fit logically into the missing portion of the figure on the left.

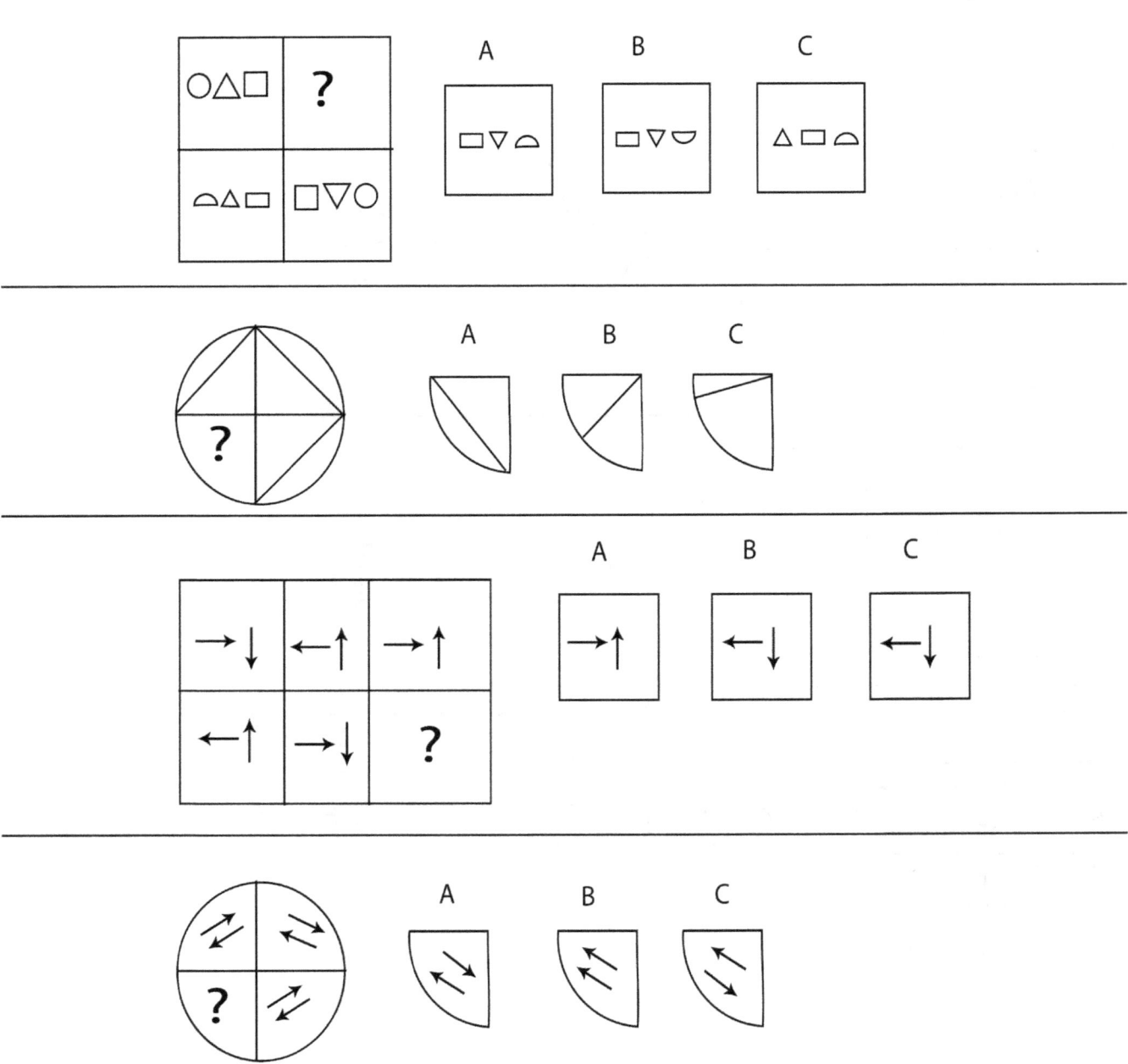

Pictorial Reasoning

PATTERN PERCEPTION - CONTINUING PATTERN

Find the correct figure from the two alternatives given, that will logically continue the pattern of figures on the left.

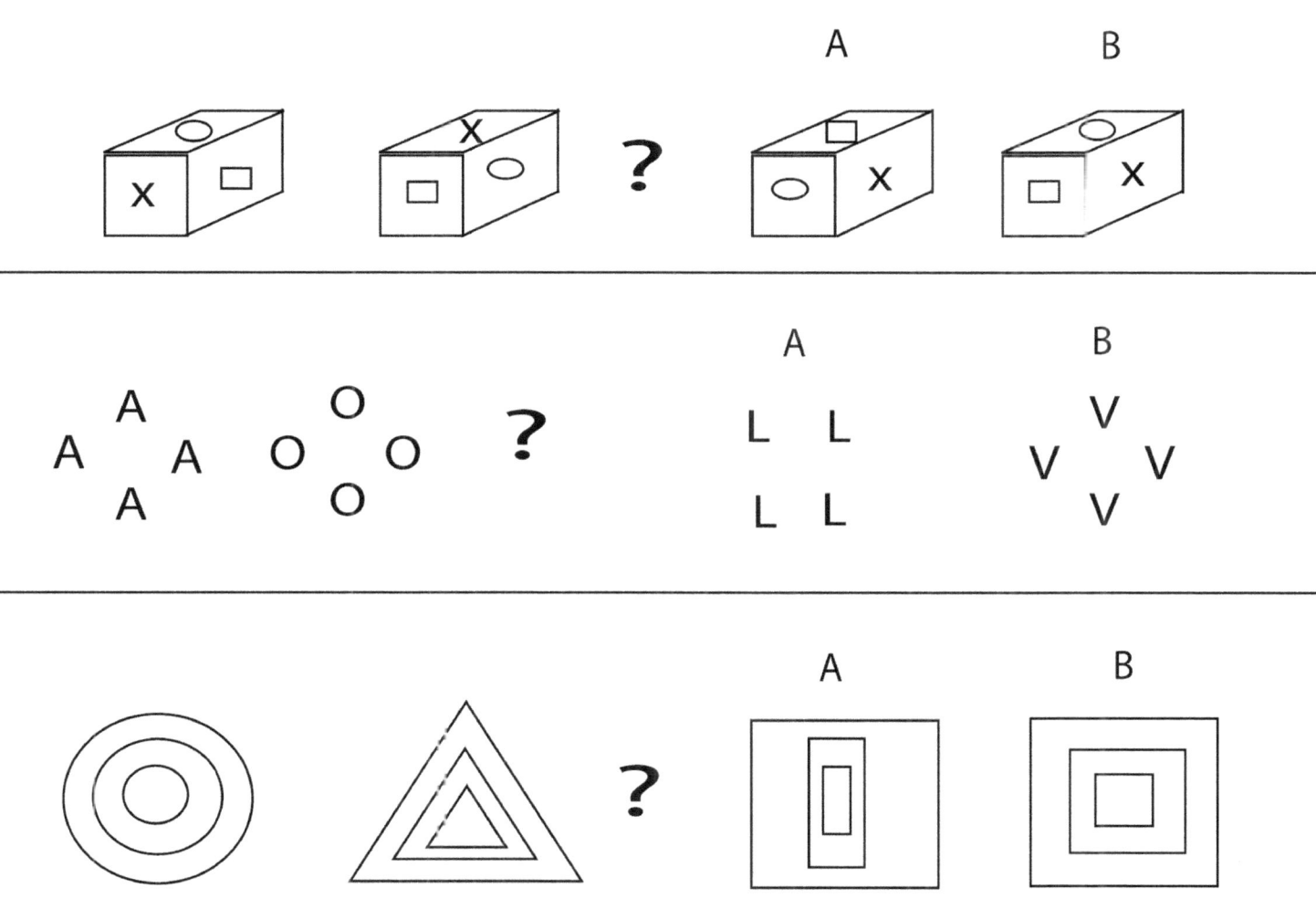

Pictorial Reasoning

FIGURE FORMATION

Find the correct figure that will be formed when the two figures on the left are combined. Either of the figures may be rotated before combining.

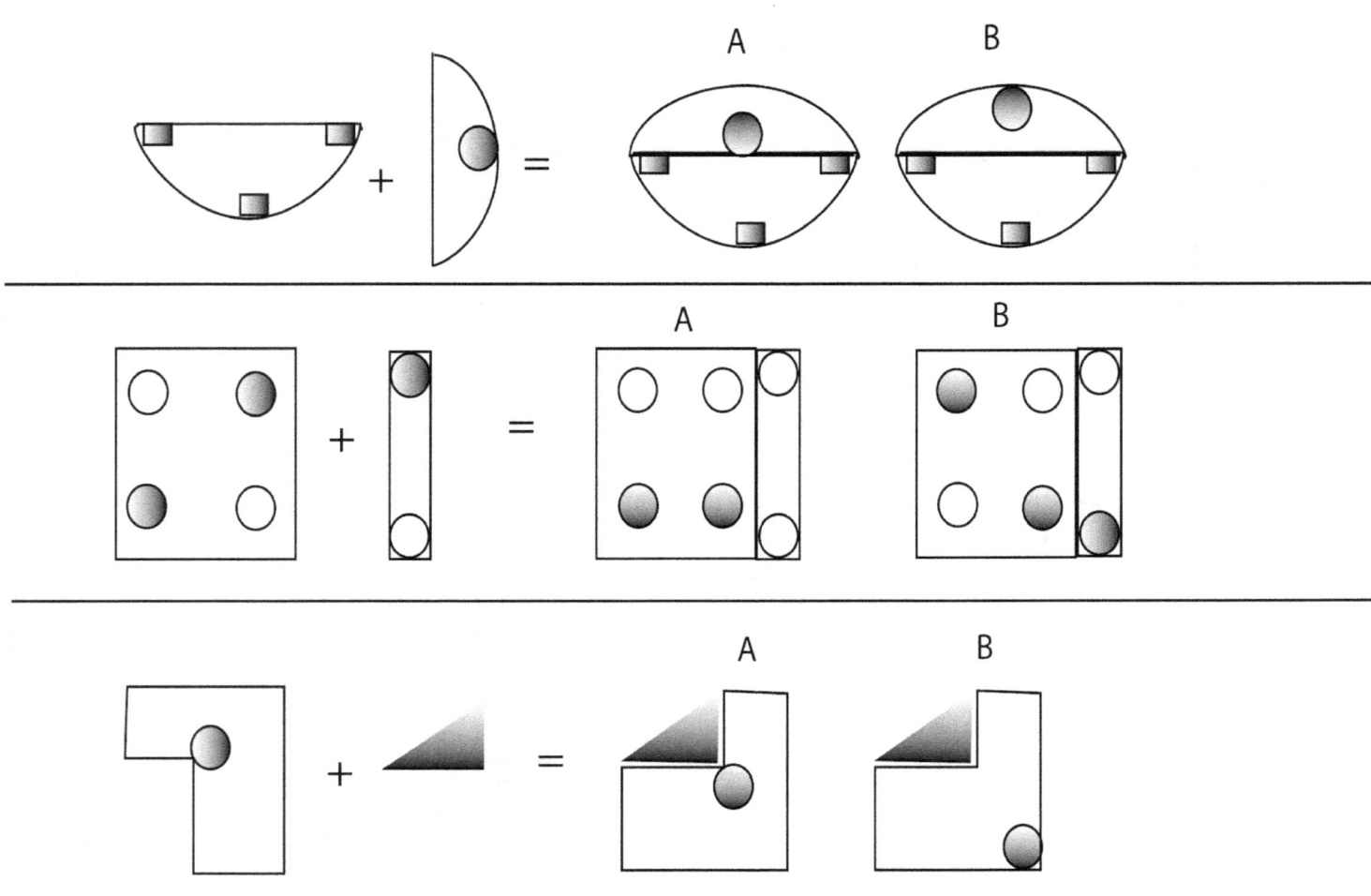

Pictorial Reasoning

PAPER FOLDING AND CUTTING

Find the correct figure that will be formed when the paper on the left is folded in the direction of the arrows, and then holes are cut in it as shown.

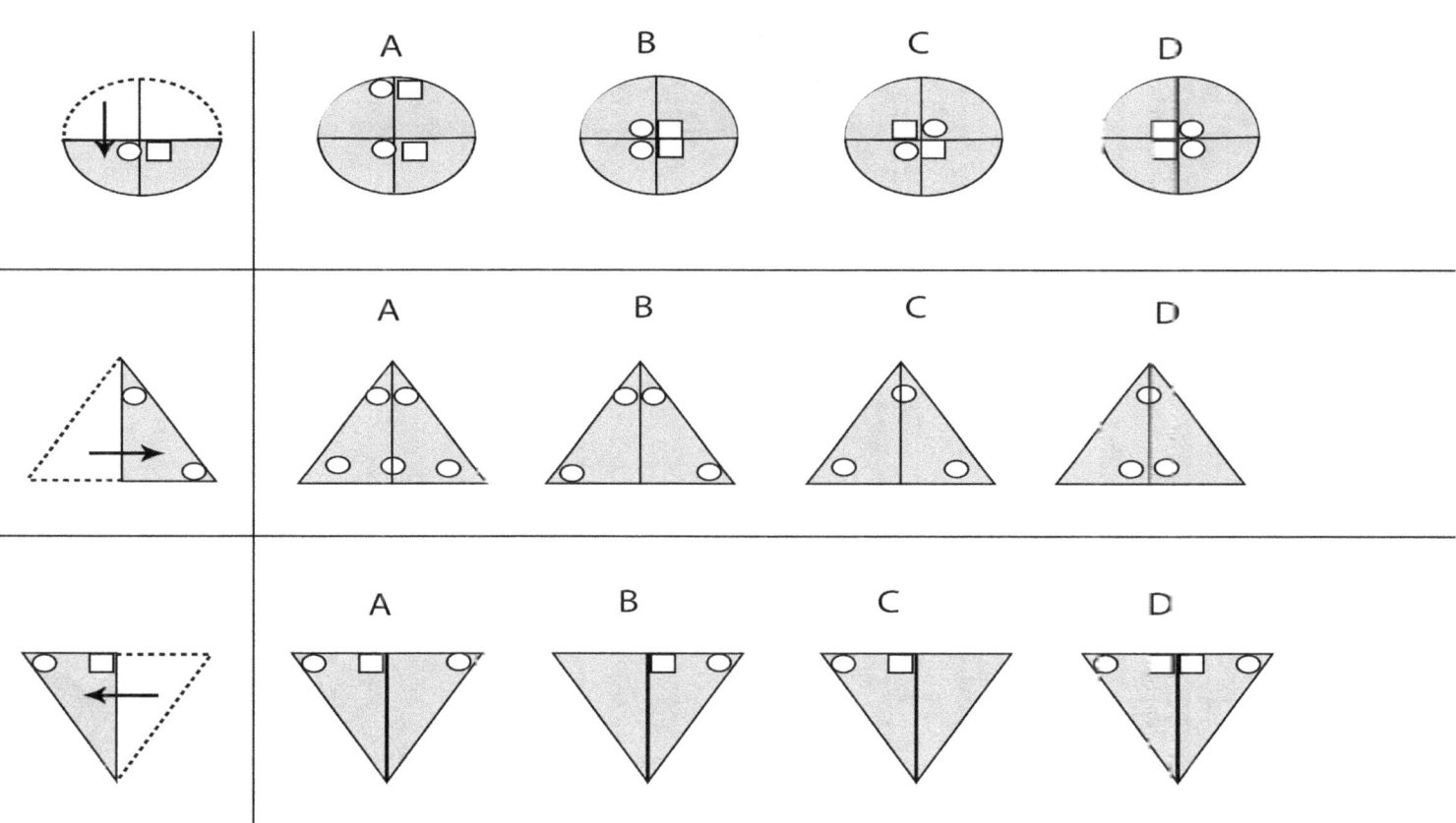

Pictorial Reasoning

FIGURE MATRIX - SIMILARITY

Three figures in the 2 x 2 matrix have similar characteristics. Find the fourth figure from the alternatives given, that is also alike.

A

B

C

A
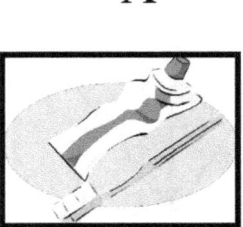

B

C
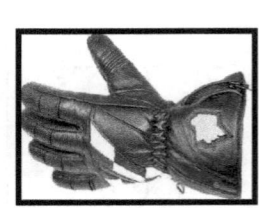

Pictorial Reasoning

FIGURE MATRIX - ANALOGY

Find the correct figure from the alternatives given that will fit in the empty box, such that the bottom two figures are related in the same way as the top two figures.

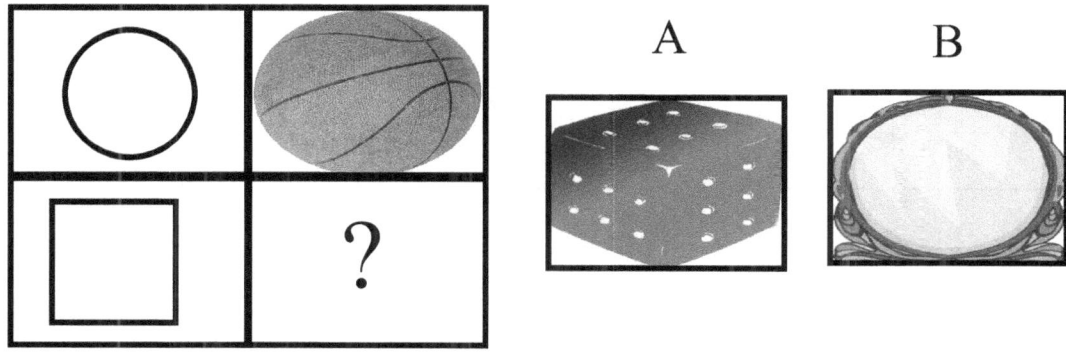

RULE DETECTION

Read the given rule in each question. Then, find the correct choice from the alternatives given that satisfies the rule.

The shaded circle moves clockwise

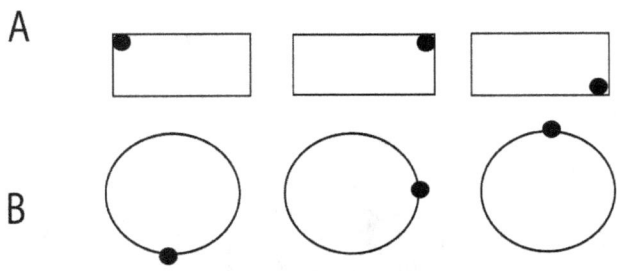

The arrows rotate clockwise

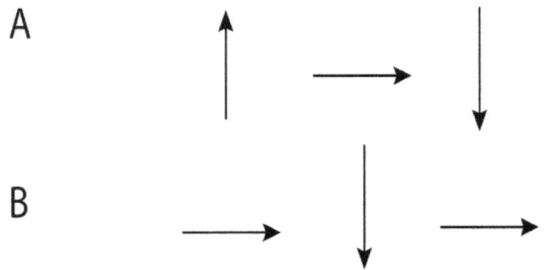

The outside figures move anticlockwise and the inside figures move clockwise

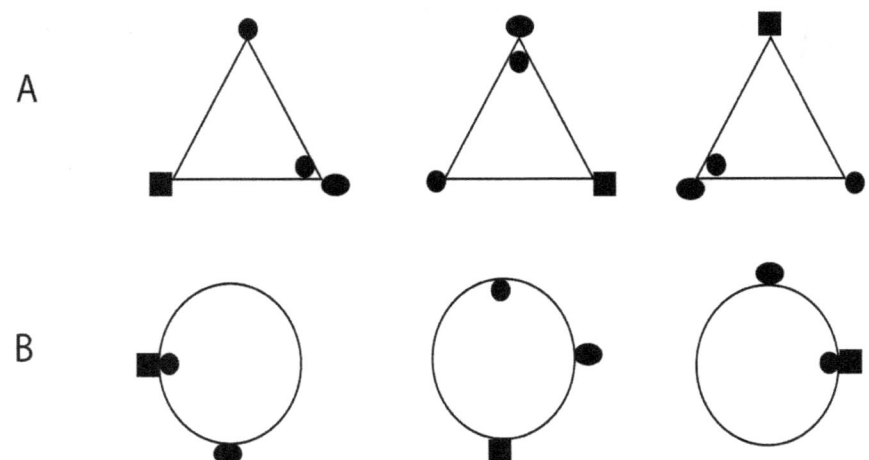

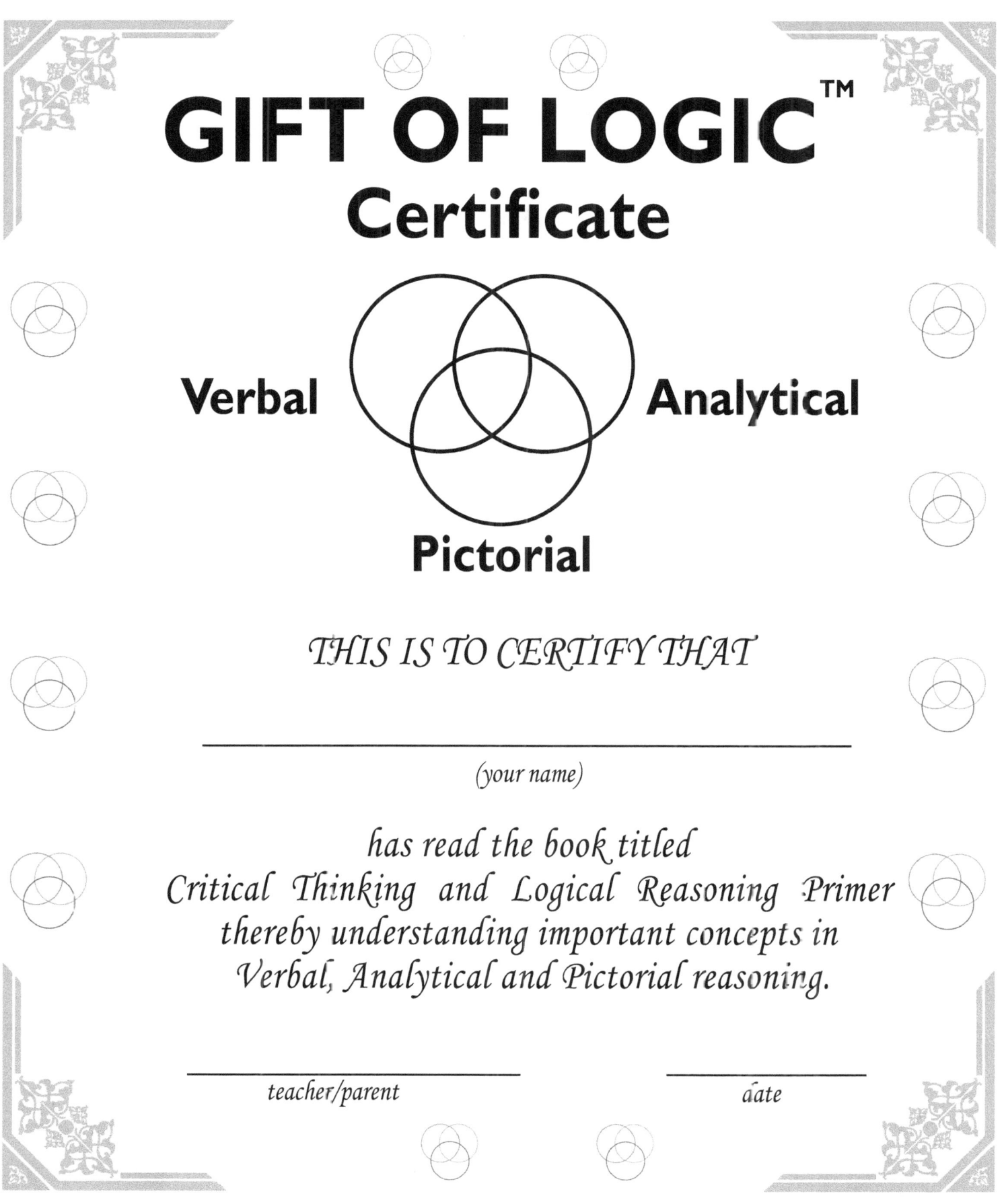

NOTES

NOTES

NOTES

NOTES

NOTES